FORENSIC ANALYTICAL
TECHNIQUES

Analytical Techniques in the Sciences (AnTS)
Series Editor: David J. Ando, Consultant, Dartford, Kent, UK

A series of open learning/distance learning books which covers all of the major analytical techniques and their application in the most important areas of physical, life and materials sciences.

Titles available in the Series

Analytical Instrumentation: Performance Characteristics and Quality
Graham Currell, University of the West of England, Bristol, UK

Fundamentals of Electroanalytical Chemistry
Paul M.S. Monk, Manchester Metropolitan University, Manchester, UK

Introduction to Environmental Analysis
Roger N. Reeve, University of Sunderland, UK

Polymer Analysis
Barbara H. Stuart, University of Technology, Sydney, Australia

Chemical Sensors and Biosensors
Brian R. Eggins, University of Ulster at Jordanstown, Northern Ireland, UK

Methods for Environmental Trace Analysis
John R. Dean, Northumbria University, Newcastle, UK

Liquid Chromatography –Mass Spectrometry: An Introduction
Robert Ardrey, University of Huddersfield, Huddersfield, UK

Analysis of Controlled Substances
Michael D. Cole, Anglia Polytechnic University, Cambridge, UK

Infrared Spectroscopy: Experimentation and Applications
Barbara H. Stuart, University of Technology, Sydney, Australia

Practical Inductively Coupled Plasma Spectroscopy
John R. Dean, Northumbria University, Newcastle, UK

Bioavailability, Bioaccessibility and Mobility of Environmental Contaminants
John R. Dean, Northumbria University, Newcastle, UK

Quality Assurance in Analytical Chemistry
Elizabeth Prichard and Vicki Barwick, Laboratory of the Government Chemist, Teddington, UK

Extraction Techniques in Analytical Sciences
John Dean, Northumbria University, Newcastle, UK

Forensic Analysis Techniques
Barbara H. Stuart, University of Technology, Sydney, Australia

FORENSIC ANALYTICAL TECHNIQUES

Barbara Stuart
University of Technology, Sydney, Australia

A John Wiley & Sons, Ltd., Publication

This edition first published 2013
© 2013 John Wiley & Sons, Ltd.

Registered office

John Wiley & Sons Ltd, The Atrium, Southern Gate, Chichester, West Sussex, PO19 8SQ, United Kingdom

For details of our global editorial offices, for customer services and for information about how to apply for permission to reuse the copyright material in this book please see our website at www.wiley.com.

Library of Congress Cataloging-in-Publication Data applied for.

A catalogue record for this book is available from the British Library.

HB ISBN: 9780470687277
PB ISBN: 9780470687284

Set in 10/12pt Times by Laserwords Private Limited, Chennai,India.

Contents

Series Preface

There has been a rapid expansion in the provision of further education in recent years, which has brought with it the need to provide more flexible methods of teaching in order to satisfy the requirements of an increasingly more diverse type of student. In this respect, the open learning approach has proved to be a valuable and effective teaching method, in particular for those students who for a variety of reasons cannot pursue full-time traditional courses. As a result, John Wiley & Sons, Ltd first published the Analytical Chemistry by Open Learning (ACOL) series of textbooks in the late 1980s. This series, which covers all of the major analytical techniques, rapidly established itself as a valuable teaching resource, providing a convenient and flexible means of studying for those people who, on account of their individual circumstances, were not able to take advantage of more conventional methods of education in this particular subject area.

Following upon the success of the ACOL series, which by its very name is predominately concerned with analytical chemistry, the Analytical Techniques in the Sciences (AnTS) series of open learning texts has now been introduced with the aim of providing a broader coverage of the many areas of science in which analytical techniques and methods are now increasingly applied. With this in mind, the AnTS series of texts seeks to provide a range of books which will cover not only the actual techniques themselves, but also those scientific disciplines which have a necessary requirement for analytical characterization methods.

Analytical instrumentation continues to increase in sophistication, and as a consequence, the range of materials that can now be almost routinely analysed has increased accordingly. Books in this series which are concerned with the techniques themselves will reflect such advances in analytical instrumentation, while at the same time providing full and detailed discussions of the fundamental concepts and theories of the particular analytical method being considered. Such books will cover a variety of techniques, including general instrumental analysis, spectroscopy, chromatography, electrophoresis, tandem techniques, electroanalytical methods, X-ray analysis and other significant topics. In addition, books in the series will include the application of analytical techniques in areas such as environmental science, the life sciences, clinical analysis, food science, forensic analysis, pharmaceutical science, conservation and archaeology, polymer science and general solid-state materials science.

Written by experts in their own particular fields, the books are presented in an easy-to-read, user-friendly style, with each chapter including both learning objectives and summaries of the subject matter being covered. The progress of the reader can be assessed by the use of frequent self-assessment questions (SAQs) and discussion questions (DQs), along with their corresponding reinforcing or remedial responses, which appear regularly throughout the texts. The books are thus eminently suitable both for self-study applications and for forming the basis of industrial company in-house training schemes. Each text also contains a large amount of supplementary material, including bibliographies, lists of acronyms and abbreviations, tables of SI units and important physical constants and, where appropriate, glossaries and references to literature sources.

It is therefore hoped that this present series of textbooks will prove to be a useful and valuable source of teaching material, both for individual students and for teachers of science courses.

Dave Ando
Dartford, UK

Preface

The public profile of forensic science has dramatically increased in recent decades and there has been a corresponding rise in the number of students undertaking forensic science degree courses at a tertiary level with the view to a professional career in this field. During this period the application of modern analytical techniques to the examination of forensic problems has expanded, particularly due to the development of small and portable cost-effective instrumentation. The availability of new techniques has led to a greater choice of tools that can be employed to analyse forensic specimens. An understanding of a broad range of analytical tools is required by today's forensic chemists and is an important aspect of their training.

The aim of this book is to provide an overview of the most commonly used analytical techniques that are of interest to forensic chemists. A clear description of how each technique works and how to prepare specimens for analysis is provided. Some techniques are widely used as standard methods, while others are yet to be established but show great potential. An explanation of how to analyse the data obtained is also provided and, for each technique, the most common forensic applications are described. There are specific issues to consider when examining forensic samples. Apart from the applicability of a technique, the issues of dealing with small quantities of material, whether a technique is non-destructive and the cost and/or portability for fieldwork must be considered.

The reader will note that there is a deliberate focus on forensic chemistry and physical evidence. Topics such as DNA analysis are intentionally not dealt with here, and forensic biology topics are well covered elsewhere. The focus here is on how to analyse samples once collected – the process of evidence collection is, of course, an important aspect of a forensic scientist's training and is an expansive topic in its own right. This book is designed for students who are undertaking a forensic chemistry based programme and require a sound knowledge of analytical techniques. Some basic tertiary mathematical and chemistry knowledge is assumed. The book will also provide a useful reference for forensic practitioners who may be interested in investigating new forms of evidence or techniques.

I hope that this book helps fill a gap in the world of forensic textbooks. Naturally many forensic science textbooks focus on the collection of evidence, but this book will provide a resource for the teaching of forensic analytical techniques. I would like to acknowledge the valuable conversations with and the data provided by a multitude of hardworking forensic scientists in police forces, law enforcement agencies and universities – not just in Australia, but worldwide. A special thanks to my colleagues and students past and present in the Centre of Forensic Science and the School of Chemistry and Forensic Science at the University of Technology, Sydney.

About the Author

Barbara Stuart, BSc (Hons), MSc (Syd), PhD (Lond), DIC, MRSC, MRACI, CChem, MANZFSS

Barbara Stuart holds the position of Associate Professor at the School of Chemistry and Forensic Science and the Centre for Forensic Science at the University of Technology, Sydney (UTS) in Australia. She received BSc (Hons) and MSc degrees in chemistry from the University of Sydney in Australia and gained her PhD at Imperial College in London in 1993. Barbara held the position of lecturer at the University of Greenwich, London before returning to Australia to take up a position at UTS in 1995. Barbara has contributed to the teaching of a broad range of topics in the chemistry, forensic and materials programmes at UTS. She also has active research interests in the fields of forensic taphonomy and archaeology, as well as in materials conservation and environmental science, and has published many papers on these topics. Barbara is also the author of five other books published by John Wiley & Sons: *Modern Infrared Spectroscopy* and *Biological Applications of Infrared Spectroscopy*, both in the ACOL series of open learning texts, and *Polymer Analysis* and *Infrared Spectroscopy: Fundamentals and Applications* in the current AnTS series of texts, as well as *Analytical Techniques in Materials Conservation*.

Acronyms, Abbreviations and Symbols

AAS	atomic absorption spectrometry
AES	atomic emission spectrometry
AFM	atomic force microscopy
ALS	alternate light source
ANN	artificial neural network
APCI	atmospheric pressure chemical ionization
ATR	attenuated total reflectance
BAC	blood alcohol concentration
BSA	N,O-bistrimethylsilylic acid
BSE	backscattered electron
BSTFA	N,O-bistrimethylsilyltrifluoroacetamide
CE	capillary electrophoresis
CI	chemical ionization
CNS	central nervous system
COHb	carboxyhaemoglobin
CZE	capillary zone electrophoresis
Δn	birefringence
DAC	diamond anvil cell
DAD	diode array detector
DART	direct analysis in real time
DC	direct current
DESI	desorption electrospray ionization
DFO	1,8-diazafluoren-9-one
DRIFT	diffuse reflectance infrared by Fourier transform
DSC	differential scanning calorimetry
DTA	differential thermal analysis
DTG	derivative thermogravimetric

ECD	electron capture detector
EDS	energy dispersive X-ray spectroscopy
EDX	energy-dispersive X-ray analysis
EDXRF	energy-dispersive X-ray fluorescence
EELS	electron energy loss spectroscopy
EI	electron ionization
ESEM	environmental scanning electron microscopy
ESI	electrospray ionization
FID	flame ionization detector
GC	gas chromotography
GHB	gamma-hydroxybutyric acid
GFAAS	graphite furnace atomic absorption spectrometry
GRIM	Glass Refractive Index Measurement
GSR	gunshot residue
Hb	haemoglobin
HCA	hierarchical clustering analysis
HDPE	high-density polyethylene
HPLC	high-performance liquid chromatography
IBA	ion beam analysis
IC	ion chromatography
ICDD	International Centre for Diffraction Data
ICP–MS	inductively coupled plasma–mass spectrometry
ILR	ignitable liquid residue
IMS	ion mobility spectrometry
IRMS	isotope ratio mass spectrometry
LC	liquid chromatography
LDA	linear discriminant analysis
LDPE	low-density polyethylene
LIBS	laser-induced breakdown spectroscopy
MDA	methylenedioxyamphetamine
MDMA	methylenedioxymethylamphetamine
MDEA	3,4-methylenedioxyethylamphetamine
MECC	micellar electrokinetic capillary chromatography (*also* MEKC)
MI	medullary index
MS	mass spectrometry
MSP	microspectrophotometry
MSTFA	N-methyl-M-trimethylsilyltrifluoroacetamide
m/z	mass-to-charge (ratio)
NMR	nuclear magnetic resonance
NOE	Nuclear Overhauser Effect
NRA	nuclear reaction analysis
PC	paper chromatography

PCA	principal component analysis
PD	physical developer
PDMS	polydimethylsiloxane
PE	polyethylene
PET	poly(ethylene terephthalate)
PETN	pentaerythritol tetranitrate
PIXE	particle-induced X-ray emission
PLM	polarizing light microscope
PMMA	poly(methyl methacrylate)
PP	polypropylene
ppb	parts per billion
ppm	parts per million
PS	polystyrene
PSA	prostate-specific antigen (test)
PVC	poly(vinyl chloride)
ρ	density
R^2	coefficient of determination
RBS	Rutherford back scattering spectrometry
RF	radiofrequency
RI	reflective index
RRS	resonance Raman spectroscopy
SAN	styrene–acrylonitrile
SAP	seminal acid phosphatase
SBR	styrene–butadiene rubber
SE	secondary electron
SEM	scanning electron microscopy
SERS	surface-enhanced Raman spectroscopy
SIMS	secondary ion mass spectrometry
SIM	selected ion monitoring
SPE	solid-phase extraction
SPME	solid-phase microextraction
TEA	thermal energy analyser
TEM	transmission electron microscopy
TFAA	trifluoroacetic anhydride
TGA	thermogravimetric analysis
THC	Δ^9-tetrahydrocannabinol
THM	thermally assisted hydrolysis methylation
TLC	thin layer chromatography
TMAH	tetramethylammonium hydroxide
TMB	tetramethylbenzidine
TOF	time-of-flight
TOF-SIMS	time-of-flight secondary ion mass spectrometry

UV	ultraviolet
UV–vis	ultraviolet–visible
VMD	vacuum metal deposition
WDXRF	wavelength-dispersive X-ray fluorescence
XRD	X-ray diffraction
XRF	X-ray fluorescence

Chapter 1

The Chemistry of Forensic Evidence

Learning Objectives

- To gain knowledge of the different classes of materials examined as forensic evidence, including fibres, paint, polymers, documents, glass, soil, explosives, firearms, arson residues, body fluids, drugs, toxicological specimens and fingerprints.
- To understand the chemistry of common types of forensic evidence.
- To gain an understanding of the analytical tools used to interpret forensic data.

1.1 Introduction

Forensic science pertains to science applied to the law, including criminal investigation, with results being presented as evidence in the law courts. Within this discipline is *forensic chemistry*, which is regarded as the application of analytical chemistry – the analysis of compounds and elements – to legal matters. Forensic chemists are asked to analyse samples to be used as evidence and draw appropriate and accurate conclusions about such evidence.

A major aspect of forensic chemistry is the analysis of physical evidence to connect a criminal to a crime. A fundamental idea in forensic science is the Locard principle, which states that every contact leaves a trace. The evidence collected from a crime scene or a victim may be matched with the evidence found on or in the possession of an individual accused of a crime.

Forensic Analytical Techniques, First Edition. Barbara Stuart.
© 2013 John Wiley & Sons, Ltd. Published 2013 by John Wiley & Sons, Ltd.

Important questions that can be asked about evidence are: what is the evidential material, and can it be linked to a crime? It is the role of a forensic chemist to identify and characterize the nature of the evidence and whether it is possible to predict the source of the material based on its chemistry. An array of laboratory techniques is used to solve these problems.

There are important considerations to be made when analysing forensic evidence as it frequently consists of very small samples and sensitive analytical techniques are required for an accurate analysis. Another challenge for forensic chemists is that specimens submitted as evidence are usually not pure substances, but are mixtures or may contain contaminants. However, the identification of a unique mixture of chemical compounds can be a positive advantage for connecting evidence to a crime scene.

1.2 Evidence Types

The type of evidence collected from a crime scene depends on the nature of a crime. *Evidence* can be categorized based on its origins and/or composition. An overview of the chemistry of commonly encountered types of evidence is provided here.

1.2.1 Polymers

Polymers are molecules consisting of a large number of repeated chemical structural units [1]. Polymers are constituents of fibre, paint and document evidence (to be described in other sections of this chapter), so they play an important role in forensic evidence. Polymers are also found in other types of evidence including packaging, adhesive tapes and vehicles. Many commercial polymers are produced, and the polymer type will depend very much on the product type and application. The majority of commercial polymers are carbon based, but some are based on inorganic structures such as silicon. Figure 1.1 illustrates the structures of regularly encountered polymers.

Although polymers are commonly referred to as plastics, this term refers to one class of polymers more correctly known as *thermoplastics*. Thermoplastics melt when heated and re-solidify when cooled. Two other classes of polymer are *thermosets*, which have cross-linked structures and decompose when heated, and *elastomers*, which are also cross-linked but with rubber-like characteristics. Some polymer materials may also be *copolymers*, where the polymer molecules consist of two or more different monomer structures. There are also polymer *blends* that are mixtures of at least two different polymers or copolymers, and composites, which are a mixture of two or more materials with a polymer being a fibre or matrix component. Additives are very often present in commercial polymers, and these can include fillers (e.g. calcium carbonate; used to improve the mechanical properties), plasticizers (e.g. esters; used to modify flexibility),

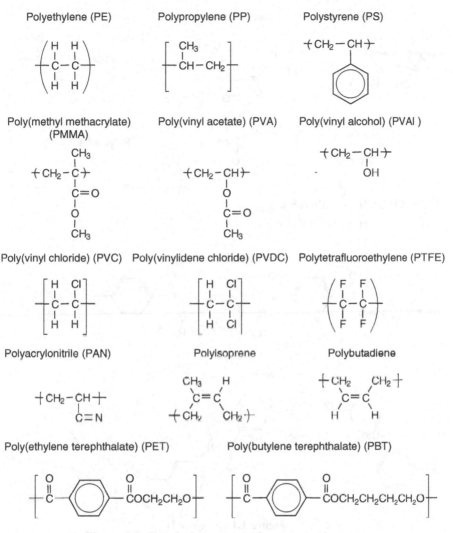

Figure 1.1 Chemical structures of common polymers.

stabilizers (e.g. carbon black; used to counteract degradation) and dyes or pigments (to impart colour).

Polymers are used as drug *packaging* or for the disposal of remains, so linking bags to a source can provide valuable information. The most commonly used polymers in packaging are polyethylene (PE) (e.g. bags and bottles), polypropylene (PP) (e.g. bottles and containers), polystyrene (PS) (e.g. containers and foam insulation) and poly(ethylene terephthalate) (PET) (e.g. food containers and bottles). Other types of bulk polymers that might be collected as evidence

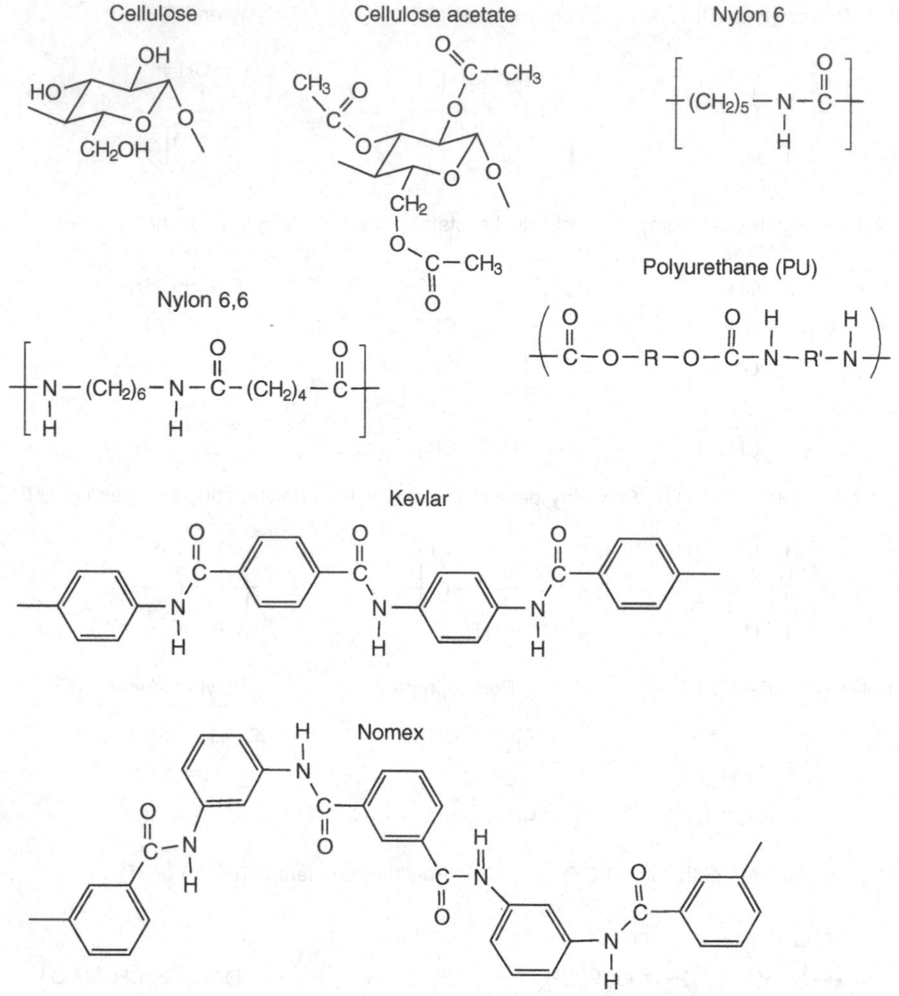

Figure 1.1 (*continued*)

include poly(vinyl chloride) (PVC), which is used to produce pipes and electrical insulation.

Adhesive tapes are also polymer based and are used in drug packaging as well as in robberies or explosive devices. Both the adhesive and tape components are manufactured using polymers. A range of tapes are commercially produced including cellotape (regenerated cellulose), packaging tape (e.g. PP) and electrical tape (e.g. PVC). The adhesives generally fall into several

Epoxy resin (bisphenol A)

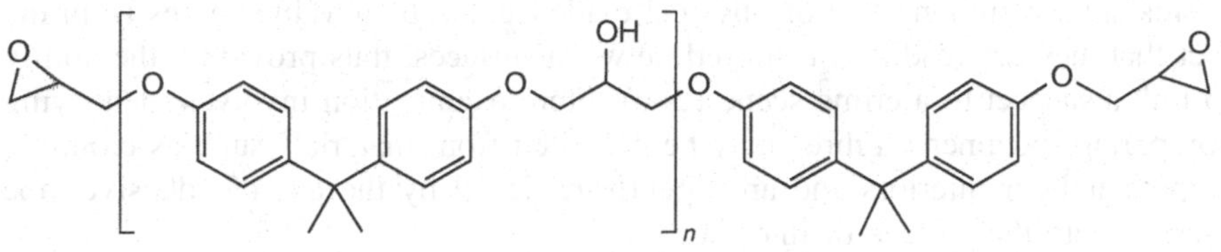

Polyester resin

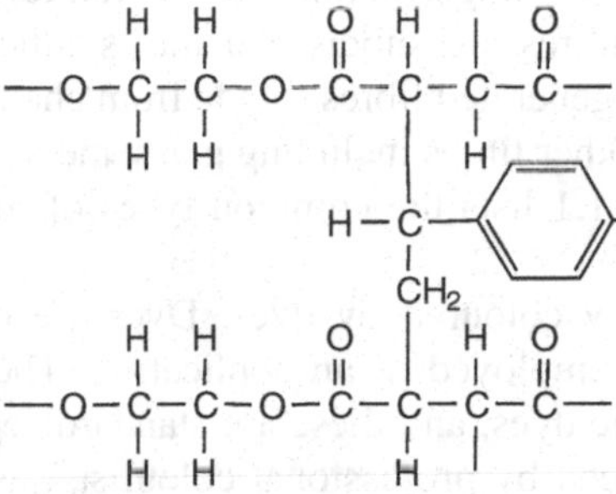

Figure 1.1 (*continued*)

main categories: acrylic (polymers derived from acrylate and methacrylate monomers), elastomer based (e.g. butadiene and isoprene) and copolymers of styrene with butadiene and isoprene.

Polymers are increasingly used in *automobile* design and can be used as evidence in accident investigation. Copolymers such as styrene–acrylonitrile (SAN) and styrene–butadiene rubber (SBR) are used in tyre manufacture and in bumper bars. Car light enclosures are produced using poly(methyl methacrylate) (PMMA).

SAQ 1.1

Suggest possible polymer types that may be identified in each of the following evidence types:

(a) plastic shopping bags

(b) adhesive tape

(c) car tyres.

1.2.2 Fibres

Fibres are a common form of physical evidence and their value comes from the fact that they are readily transferred between surfaces, thus providing the ability to link a suspect to a crime scene [2–6]. Fibre examination involves identifying comparing specimens. Fibres may be collected from materials such as clothing, carpets and car interiors and are directly retrieved by the use of adhesive tape pressed onto the surface of interest.

An extensive range of fibres are classified into two groups: natural and man-made. Natural fibres can derive from animal sources (e.g. hair, wool or silk), plant sources (e.g. cotton, linen, hemp, jute or ramie) or mineral sources (e.g. asbestos). Man-made fibres are subdivided into synthetic fibres (made by synthetic polymers) and regenerated fibres (made from chemically modified naturally occurring polymers). Other fibres including glass, metal or carbon fibres may also be encountered. Table 1.1 lists the common types of man-made fibres and their compositions.

Fibres are principally coloured by dyes. Dyes are colouring agents that are soluble in the solvent employed in an application. There is an extensive range of commercial synthetic dyes, and these are standardized by a Colour Index (CI) number that is referenced by professional colourists' associations [7]. Dyes can be classified by their structure type or the method of application. Some common structural types used as dyes include azos (containing an —N=N— structure),

Table 1.1 Common man-made fibres

Fibre type	Composition
Regenerated	
acetate	cellulose acetate
triacetate	cellulose triacetate
viscose rayon	regenerated cellulose (precipitated from acidic solution)
lyocell	regenerated cellulose (precipitated from organic solution)
azlon	regenerated protein
Synthetic	
acrylic	> 85 % acrylonitrile
modacrylic	35–85% acrylonitrile
polyester	PET, PBT
nylon	nylon 6, nylon 6,6, nylon 11
aramid	Kevlar, Nomex
urethane	polyurethane (e.g. Spandex and Lycra)
olefin	PE, PP
chlorofibre	PVC, PVDC
vinyon	PVC
fluorofibre	PTFE
vinal	PVAl

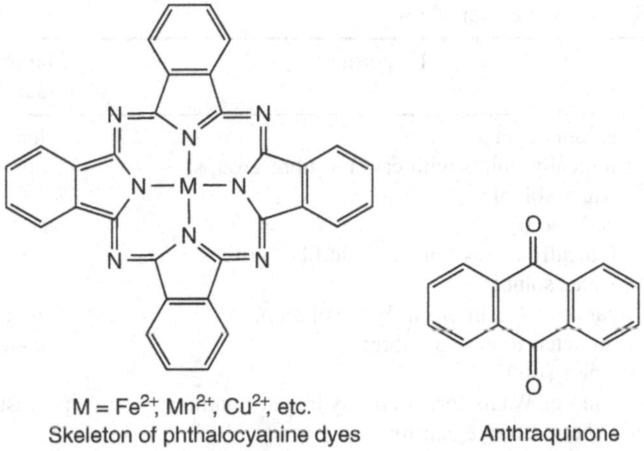

M = Fe^{2+}, Mn^{2+}, Cu^{2+}, etc.
Skeleton of phthalocyanine dyes   Anthraquinone

Figure 1.2 Base structures of phthalocyanines and anthraquinones.

phthalocyanines (Figure 1.2), anthraquinones (Figure 1.2), carbonyls and polyme-thines. Dye groups are also classed by the method of application, and Table 1.2 summarizes the properties of the main dye classes.

SAQ 1.2

A blue polyester fibre is collected as evidence. What is the likely class of dye used to colour this fibre? What is the possible nature of the bonding between the dye and the fibre in this case?

1.2.3 Paint

Paint is used to cover a range of surfaces, but the types that are of most interest in forensic science are automotive and architectural paints [8–11]. Automotive paint is often collected as evidence for car accidents or incidents. Architectural paints can be transferred during robberies, for instance. Due to the enormous range of commercial paints available with different compositions, paint is a valuable form of evidence. Additionally, because paint specimens are commonly multi-layered, more discriminating information can be provided.

Paints are composed of pigments, binders, solvents and additives. *Pigments* are colouring agents that are suspended particles in a solvent. There is an array of commercial pigments available used to impart colour, and some common paint pigments are listed in Table 1.3. Metals can also be added to certain paint types, such as metallic automobile paints. Pigments may be inorganic or organic compounds and synthetic or naturally occurring in origin. The *binder* provides the supporting medium for the pigment and additives, and enables a film to be formed

Table 1.2 Dye classes used for fibres

Dye class	Properties	Common fibre substrates
Acid	anionic dye; ionically bonds with cationic fibre groups; water soluble	nylons, PP, wool, silk
Basic	cationic dye; ionically bonds with anionic fibre groups; water soluble	acrylic, PAN
Direct	cationic dye in electrolytic solution; attracted to anionic fibre; water soluble	acetate, rayon, cotton
Disperse	van der Waals forces and hydrogen bonding between dye and fibre; water insoluble	polyester, acetate
Reactive	covalent bonding between dye and fibre; water soluble	cotton, wool, cellulose-based
Vat	reductant used to solubilize dye; oxidized in fibre; water insoluble	cellulose-based
Sulfur	reductant used to solubilize sulfur-based dye; oxidized in fibre; water insoluble	cellulose-based

Table 1.3 Common paint pigments and extenders

Compound	Properties
aluminium silicates (clays)	extenders
anatase (TiO_2)	white pigment
barium chromate ($BaCrO_4$)	yellow pigment
barium sulfate ($BaSO_4$)	white pigment
cadmium sulfides (CdS)	yellow, red and orange pigments
calcium sulfates ($CaSO_4 x H_2O$)	extenders
calcium carbonates ($CaCO_3$)	extenders
cerulean blue (CoO, SnO_2)	blue pigment
chrome reds ($PbCrO_4$, $Pb(OH)_2$)	red pigments
chrome yellows ($PbCrO_4$, $PbSO_4$)	yellow pigments
chromium oxide (Cr_2O_3)	green pigment
cobalt blue (CoO, Al_2O_3)	blue pigment
magnesium carbonate ($MgCO_3$)	extender
magnesium silicate ($Mg_3Si_4O_{10}(OH)_2$)	extender
ochres (Fe_2O_3)	yellow, red and brown pigments
Prussian blue ($Fe_4(Fe(CN)_6)_3$)	blue pigment
red lead (Pb_3O_4)	red pigment
rutile (TiO_2)	white pigment
viridian ($Cr_2O_3.2H_2O$)	blue-green pigment

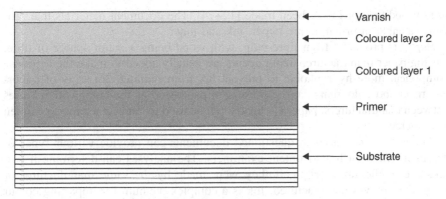

Figure 1.3 Paint layer structure.

when the paint dries. Alkyds and acrylic resins are common binders. Alkyd resins are cross-linked or linear polymers produced from alcohols and acids, and are usually modified with oil. There is a broad range of organic *solvents* used in paint, and the use of a solvent assists in the formation of a uniform film. The drying process is usually a combination of solvent evaporation and a polymerization process. Drying oils, such as linseed oil, are sometimes present to produce films resulting from the oxidation of the oil. Common additives that can be present in paint include extenders and plasticizers. *Extenders* are often low-cost pigments used to reduce the cost of the paint. *Plasticizers* are compounds, such as phthalate esters, that are used to increase flexibility.

Paint can be applied to nonporous surfaces, such as metals, or to porous surfaces, such as wood. A layered system is usually found, with primer, coloured and varnish (or clearcoat) layers regularly used. The primer is used to prepare the substrate for the application of paint; the varnish protects the paint layer and also provides gloss and/or colour improvement. The number and nature of layers will vary between applications. Figure 1.3 illustrates a typical paint layer structure.

Automotive paint is a common type of paint evidence and can have a complex layering system that usually consists of at least four coatings. The first layer is an electrocoat polymer consisting of epoxy resin that is electroplated onto steel to provide corrosion resistance. The second layer is usually a primer surface composed of epoxy–polyester resin or urethane and pigments to provide an even surface. The next layer is the basecoat (or colourcoat), which determines the paint colour and often has an acrylic-based binder. The final clearcoat is often acrylic based or urethane based.

1.2.4 Documents

Questioned documents are a well-known form of evidence. Analysis of the chemical composition of documents provides information about the origins and whether or not modifications have been made [12–14]. The document materials that provide valuable information are paper, ink and toner.

Paper is produced from fibre pulp, with wood being a major source of fibre. During paper manufacture, sizing agents are usually added to make the cellulose component more hydrophobic to prevent ink from running. An array of minerals, resin and colourants can be present in paper and as the composition varies between manufacturers, paper chemistry can be used to connect a paper specimen to a source.

Analysis of *inks* used in questioned documents can provide various types of useful information for document examiners. The analyses can determine if two inks are of the same origin, if they were made by the same manufacturer or when the ink was manufactured. Ink is a complex medium for imparting colour and can include dyes, pigments, solvents, resins and lubricants. Modern writing instruments fall into two main categories: ballpoint pens and non-ballpoint pens. Ballpoint pens contain oil-based inks and use dyes as colourants. Non-ballpoint pens contain water-based inks and include fountain, rollerball and gel pens.

Toner is in widespread use in printed documents, and the variation in ingredients means that analyses can be used to determine if documents are printed or copied with the same toner, the source of a printed or copied document or if alterations have been made to a document. Dry toners can consist of pigments (e.g. carbon black), polymer binders (e.g. styrene–acrylate copolymer, styrene–butadiene copolymer, polyester resin and epoxy resin) and additives (e.g. ferrite and magnetite).

1.2.5 Glass

Glass is a hard, brittle material that can be broken into small fragments and is a common form of evidence [15, 16]. Glass may be recovered from the clothing of a suspect, for instance, or be found at the scene of car accidents. Glass is produced when silica (SiO_2) (most commonly) and other metal oxides are melted at a high temperature and then quickly cooled to produce an amorphous structure. Fluxes, such as Na_2O or K_2O, are added to lower the melting temperature of the mixture, and work by disrupting the Si—O network produced by silica. The oxygen atoms become negatively charged and loosely hold the monovalent cations within the network (Figure 1.4). As the bonding is weak, cations can migrate out of the network in water, so a stabilizer such as lime (CaO) or magnesia (MgO) is added to make the glass water resistant. Stabilizers are divalent and so are held more tightly in the glass structure. A number of trace elements are also present in glass, and these can be natural impurities or deliberately added to produce colour.

There is a vast number of glass formulations, but glass can be classified by several main compositional classes. *Soda–lime glass*, also known as soda–lime–silicate glass, is the most common class encountered in forensic evidence. Soda–lime glass is formed from SiO_2 with Na_2O and K_2O added as fluxes and CaO, Al_2O_3 and MgO added as stabilizers. This is the most common

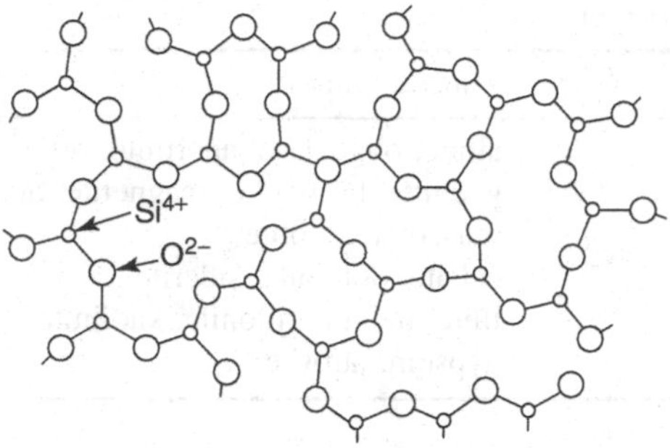

Figure 1.4 Silica glass structure.

composition of glass used for flat glass, bottles or containers. Borosilicate glass is a soda–lime glass containing more than 5% boric acid (B_2O_3) and is used in automobile headlights and heat-resistant glass such as Pyrex.

Different manufacturing processes are used to produce glass for its various applications. Float glass is produced by layering molten glass on a bath of molten tin in an inert atmosphere to produce a flat surface, and this type of glass is widely used in buildings and vehicles. Toughened or tempered glass is stronger than regular glass due to the rapid heating and cooling process during manufacturing to introduce stress. Toughened glass is used for safety glass and in vehicles. Glass can also be laminated when multiple layers of glass are bonded with a plastic film, and this form is used for windscreens and security glass. Glass for containers is produced using a blow-moulding process.

SAQ 1.3

What manufacturing process is used to produce the following glass types?

(a) car windscreen.

(b) wine bottle.

1.2.6 Soil

Soil is used as evidence as it may be found adhering to shoes, clothing or vehicles and can link a suspect to a crime scene [17–19]. The most common forensic soil examination involves a comparison of soil samples from known locations with soil collected as evidence. Soil is a complex mixture of minerals and organic matter, as well as man-made materials such as glass. Minerals form the inorganic component of soil, which is the major constituent. The inorganic component of soils results from the weathering of rocks and consists of minerals of various sizes

Table 1.4 Common soil minerals

Mineral class	Mineral names
feldspars	albite, oligoclase, anorthite, orthoclase, microcline
oxides	goethite, haematite, magnetite, quartz, anatase, gibbsite
micas	muscovite, biotite
carbonates	calcite, dolomite, siderite
clay minerals	illite, montmorillonite, kaolinite, halloysite, smectites
sulfates	gypsum, jarosite

and compositions. The inorganic component can be categorized by size: sand (2–0.02 mm), silt (0.02–0.002 mm) and clay (< 0.002 mm). Certain minerals are more commonly found in soil (these are shown in Table 1.4), but the exact composition will vary between locations. Organic matter is a minor component of most soils (< 5 %) and can derive from seeds, pollen and decomposing plants and animals. Due to the variation of soil composition over relatively small distances, it is important to check for variability of composition in the vicinity of the source.

1.2.7 Explosives

An explosion is the result of a chemical or mechanical process that results in the rapid expansion of gases. The chemical identification of explosives is used to link a suspect to an explosion or an attempted explosive incident [20–22]. Explosive evidence may be collected pre-blast, for instance from a disarmed device, or post-blast from debris collected from the scene of an explosion. Explosives are classified as low or high based on the speed of decomposition. Low explosives are ignited by a spark or a flame in a confined environment, and propagation occurs from particle to particle. High explosives generally require an initiator for detonation to occur and produce high-pressure shock waves. High explosives are generally more powerful than low explosives.

Explosives can be distinguished by their chemical structure. Figure 1.5 illustrates the structures of common explosives. Nitro compounds containing three or more nitro groups on one benzene ring and some compounds with two nitro groups are used as explosives. Nitric esters which contain a nitroxy group ($-C-O-NO_2$) and nitramines ($-C-N-NO_2$) are also compound classes used for the production of explosives. The salts of nitric, chloric or perchloric acids are also used as explosives or in explosive mixtures. Azides are another class of compounds used.

SAQ 1.4

Which of the explosives shown in Figure 1.5 are nitramines?

Low explosives

Lead azide   Ammonium nitrate

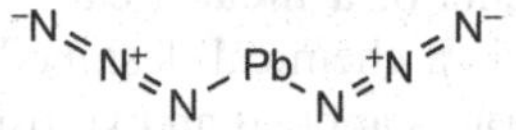

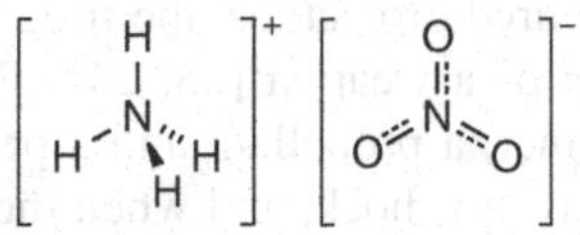

High explosives

Nitroglycerine (NG)   Picric acid 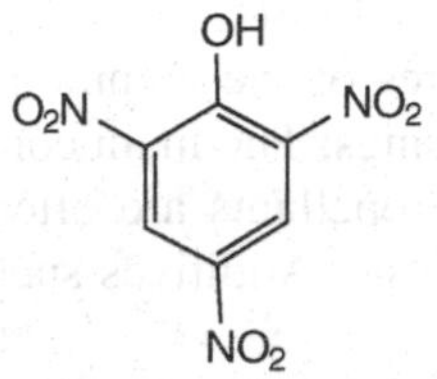   2,4,6-trinitrotoluene (TNT)

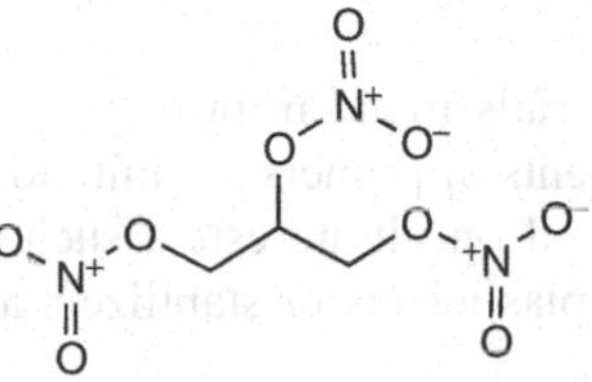

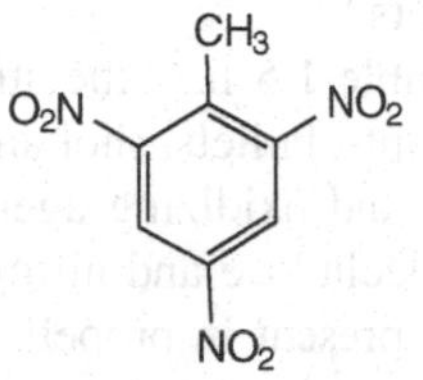

1,3,5,7-tetranitro-1,3,5,7-tetrazacyclooctane   1,3,5-trinitro-1,3,5-triazacyclohexane
(HMX)   (cyclotrimethylenetrinitramine) (RDX)

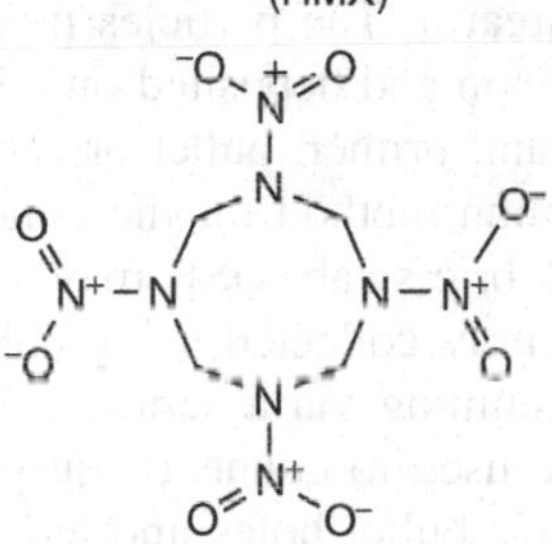

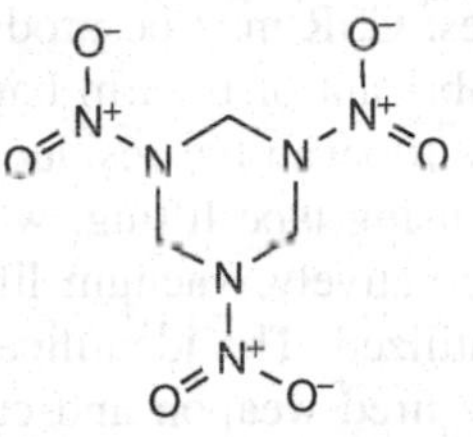

Pentaerythritol tetranitrate (PETN) 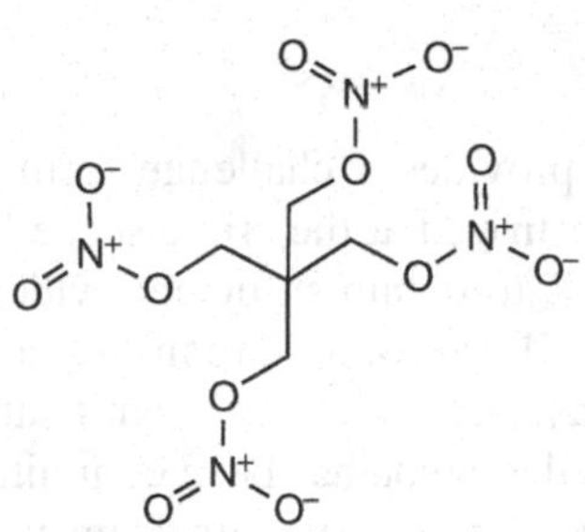   2,4,6-trinitrophenylmethylnitramine (tetryl)

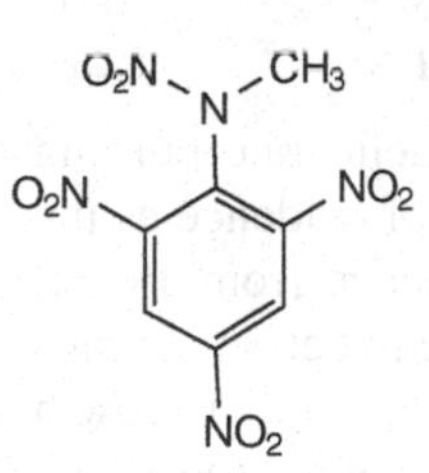

Figure 1.5 Common explosive structures.

1.2.8 Firearms

In the field of firearm examination, information about a crime can be obtained from the fired projectiles, the fired weapon itself and the residues that result from the firing of a weapon [23, 24]. The general components of a firearm cartridge are a primer, a propellant and a projectile. The primer is a chemical designed to be sensitive to shock, and when the trigger of the weapon is applied an explosion of the primer is initiated. The explosive primer causes the propellant to combust, and high temperature and pressure gases are produced. The increased gas pressure allows the projectile to travel at high speed down the barrel. For shot cartridges, there is an additional layer of material (a wad) to separate the shot from the shot pellets.

Table 1.5 lists the structures of the common materials found in primers, propellants, bullets, shot and casings. The main components of primers are initiators, fuel and oxidizing agents. Propellants are often based on nitrate esters such as nitrocellulose and nitroglycerine. Additives such as plasticizers or stabilizers are also present in propellants.

Forensic chemistry comes into play for the identification of gunshot residue (GSR). GSR results from the cooling and condensation processes of the gases of the combustion reactions that occur within a firearm. The particles produced can be expelled through various openings in a weapon and deposited on all sorts of surfaces. GSR may be produced by the propellant, primer, bullet or cartridge casing, lubricant or firearm barrel. The most common method for the collection of GSR is to swab the residue with a moistened fabric swab. GSR may also be collected using tape lifting, where residues are simply collected using adhesive tape. Alternatively, vacuum lifting, involving vacuuming via a series of filters, may be utilized. The identification of GSR can be used to connect a suspect to a recently fired weapon and can be used to identify bullet holes and aid in the determination of firing distance.

1.2.9 Arson

Arson is the deliberate burning of property and provides a challenge from the point of view of evidence as this can be largely destroyed at the crime scene [21, 25]. The challenge from a chemistry perspective is to obtain sufficient evidence from the scene to determine the cause of the arson. The agent used to initiate a fire or to increase the rate of growth of fire is known as accelerant. The identification of an accelerant at the scene of a fire demonstrates that arson has been committed. After the fire is extinguished, the residues of an accelerant remain from hours to days after the incident, so an expedient collection of evidence is required. Porous materials, such as carpets and soft furnishings, that can potentially contain flammable residues can be collected at the scene and stored in airtight containers.

The most common means of arson is the use of an ignitable liquid with ignition via a flame. The majority of arson cases are started with petroleum-based

Table 1.5 Chemical composition of common firearm projectiles

Component type	Substance	Structure
primer initiator	lead styphanate	
	diazodinitrophenol (dinol)	
primer fuel	antimony sulfide	SbS
primer oxidant	barium nitrate	$Ba(NO_3)_2$
primer cap	brass	Cu—Zn
	copper	Cu
propellant	nitrocellulose	
	nitroglycerine	
propellant plasticizer	dibutylphthalate	
	camphor	

(*continued overleaf*)

Table 1.5 (*continued*)

Component type	Substance	Structure
propellant stabilizer	diphenylamine	(see structure)
bullets or shot	lead	Pb
	lead–antimony	Pb—Sb
bullet jacket	copper	Cu
	copper–zinc	Cu—Zn
	copper–tin	Cu—Sn
shot cartridge	polyethylene	$-(CH_2\text{-}CH_2-)_n-$

accelerants, such as petrol (gasoline) or kerosene. Accelerants can be broadly grouped according to their physical state: gases (e.g. propane, butane and natural gas), liquids (gasoline and kerosene) and solids (flash powder and gunpowder). Although an ignitable liquid is consumed in the fire, ignitable liquid residues (ILRs) can be detected in fire debris.

1.2.10 Body Fluids

Body fluids, including blood, semen and saliva, can provide valuable physical evidence. The presence and distribution of the dried stains of body fluids can substantiate a crime [26, 27]. *Blood* is a biological fluid comprising plasma (containing proteins, lipids and electrolytes), erythrocytes (red blood cells), leukocytes (white blood cells) and thrombocytes (platelets). *Semen* is a suspension of spermatozoa cells in seminal fluid excreted by exocrine glands. The fluid can contain sugar, citric acid, calcium and proteins. *Saliva* is the secretion produced by the salivary glands and contains enzymes including amylase.

1.2.11 Drugs and Toxicology

Drug abuse has an enormous social cost, and drugs are a major source of forensic evidence [28–30]. There are many drugs of interest, each with different chemical properties to address. Forensic drug analysis can involve the examination of bulk or trace amounts of illicit or controlled substances. *Forensic toxicology* also involves the analysis of drugs and poisons in the body [31, 32]. Toxicological samples may be collected ante-mortem or post-mortem. Various body fluids including blood, urine, saliva and sweat and tissues, organs and hair can be used for drug analysis.

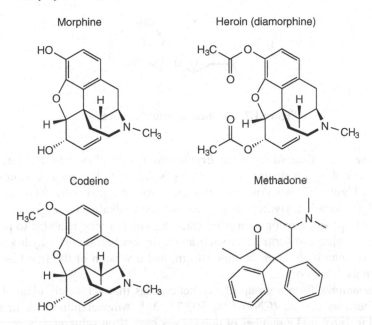

Figure 1.6 Chemical structures of main opioids.

Opiates, naturally occurring alkaloid analgesics, are obtained from the opium poppy (*Papaver somniferum*) and have been in use for thousands of years [28, 29, 33–34]. Opioid is the term used to describe the natural and semi-synthetic alkaloids prepared from opium. The structures of the main opioids are shown in Figure 1.6. Morphine is the principal opium alkaloid. There are several purification methods for morphine including the addition of calcium hydroxide or a pH adjustment approach. Codeine is also present in opium. Heroin is a semi-synthetic derivative of morphine that was first synthesized in the 19th century. To produce heroin, morphine is acetylated to produce diamorphine. Opiates are commonly found in a hydrochloride salt form. Given that raw opium contains a mixture of compounds, the final product will contain a variety of products. Heroin is also diluted with sugars, caffeine or barbiturates. Fully synthetic opioids without morphine structures have also been developed, including methadone.

SAQ 1.5

Heroin (diamorphine) can be produced by refluxing morphine in acetic anhydride. Given that codeine is also present in the opium poppy, what will be the by-product formed as a result of the reaction with acetic anhydride?

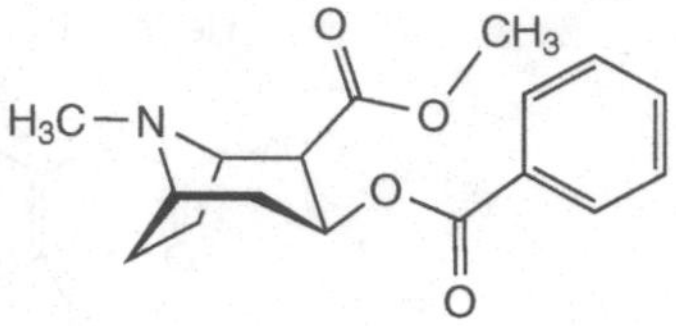

Figure 1.7 Chemical structure of cocaine.

Cocaine is an alkaloid from the *Erythroxylon coca* plants and has been recognized as a drug for thousands of years [28, 29, 33]. The cocaine structure is shown in Figure 1.7. Alkaloids are extracted from the leaves by solvent extraction and precipitation. Hydrolysis is used to convert alkaloids to ecgonine, which is treated to produce ecgonine methyl ester, to which a base is added to produce cocaine. Cocaine can form salts, such as the hydrochloride. The hydrochloride form of cocaine is the most common form, and a version of the free base form is known as crack cocaine.

Amphetamines form a group of synthetic drugs that act as stimulants for the central nervous system (CNS) [28, 30, 33, 35]. Amphetamine was first synthesized in 1887, and a number of derivatives have been subsequently produced. Amphetamines can be found in powder, liquid, crystal, tablet or capsule form. The commonly encountered derivatives include methamphetamine, 3,4-methylenedioxyamphetamine (MDA), methylenedioxymethylamphetamine (MDMA, also called Ecstasy) and 3,4-methylenedioxyethylamphetamine (MDEA). The structures of amphetamine and common derivatives are shown in Figure 1.8. Sulfate salts are commonly found forms of amphetamines. There are many synthetic routes used in clandestine laboratories to produce amphetamines. For instance, common synthetic methods for methamphetamine involve reduction reactions that add an amine group to a phenethylamine structure (known as reductive amination). Ephedrine and pseudoephedrine (Figure 1.8) have been common precursors in the reactions, and the Leuckart reaction and the Birch method are well-known synthetic routes.

Hallucinogens form a group of drugs that alter the perception of reality [28, 29, 33]. Figure 1.9 shows the structures of commonly encountered hallucinogenic drugs. Cannabis (also referred to as marijuana or hashish) is a widely used drug and is obtained from the *Cannabis sativa* plant. Cannabis is usually in the form of plant materials or resin, and the active ingredient is Δ^9-tetrahydrocannabinol (THC). Lysergic acid diethylamide (LSD) is a potent hallucinogen that has been known since the 1930s. LSD is a semi-synthetic drug, with lysergic acid being a naturally occurring ergot alkaloid. LSD is a colourless liquid and is usually found impregnated in a substance such as paper. Psilocybin and psilocin are found in a number of species of mushrooms (i.e. 'magic mushrooms'). Mescaline (3,4,5-trimethoxyphenethylamine) is also a potent hallucinogen obtained from a natural

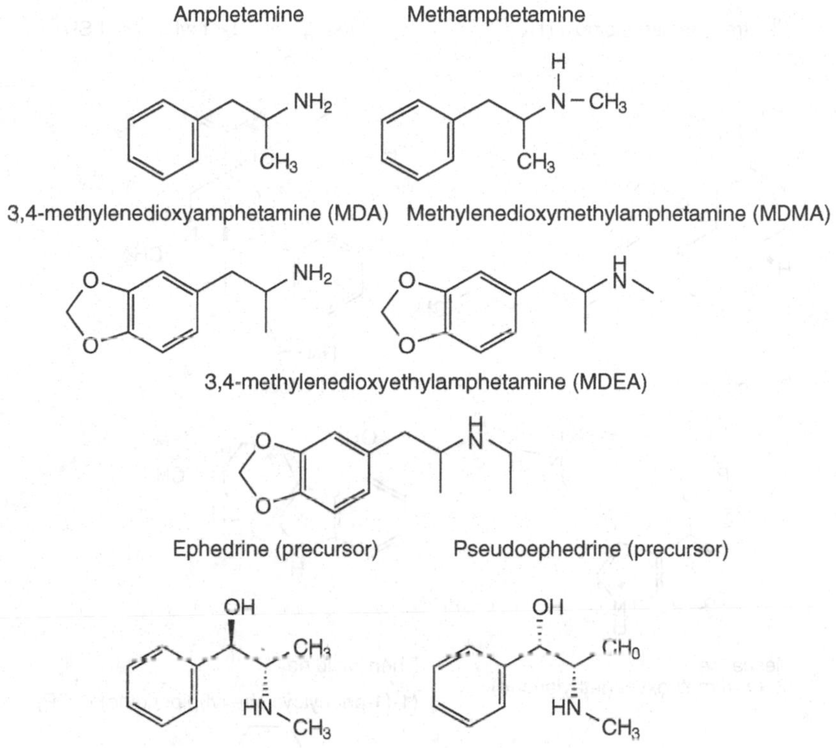

Amphetamine   Methamphetamine

3,4-methylenedioxyamphetamine (MDA)   Methylenedioxymethylamphetamine (MDMA)

3,4-methylenedioxyethylamphetamine (MDEA)

Ephedrine (precursor)   Pseudoephedrine (precursor)

Figure 1.8 Chemical structures of amphetamines.

source, an American cactus species (commonly known as peyote). Phencyclidine (PCP) (1-(1-phenylcyclohexyl)piperidine) was first synthesized as an anaesthetic in the 1950s, but was adopted as a recreational drug in the 1960s for its hallucinogenic properties. A structurally related anaesthetic, ketamine, has also been more recently adopted for its hallucinogenic properties.

Alcohol commonly refers to ethanol, which is found in alcoholic drinks, and is the most common toxic substance encountered in forensic toxicology [28, 36–38]. Alcohol acts as a depressant. Excessive consumption of alcohol is associated with many accidents, so the ability to measure the concentration of alcohol in body fluids is of major forensic importance. For ante-mortem specimens, alcohol levels in breath, blood and urine are used. Various body fluids, such as bile, are collected post-mortem in addition to blood and urine for alcohol analysis. Alcohol is initially absorbed via the stomach and into the bloodstream. Elimination of alcohol is accomplished via oxidation and excretion. Oxidation occurs in the liver, where the enzyme alcohol dehydrogenase converts the alcohol into acetaldehyde and then acetic acid, which is oxidized to carbon dioxide (CO_2)

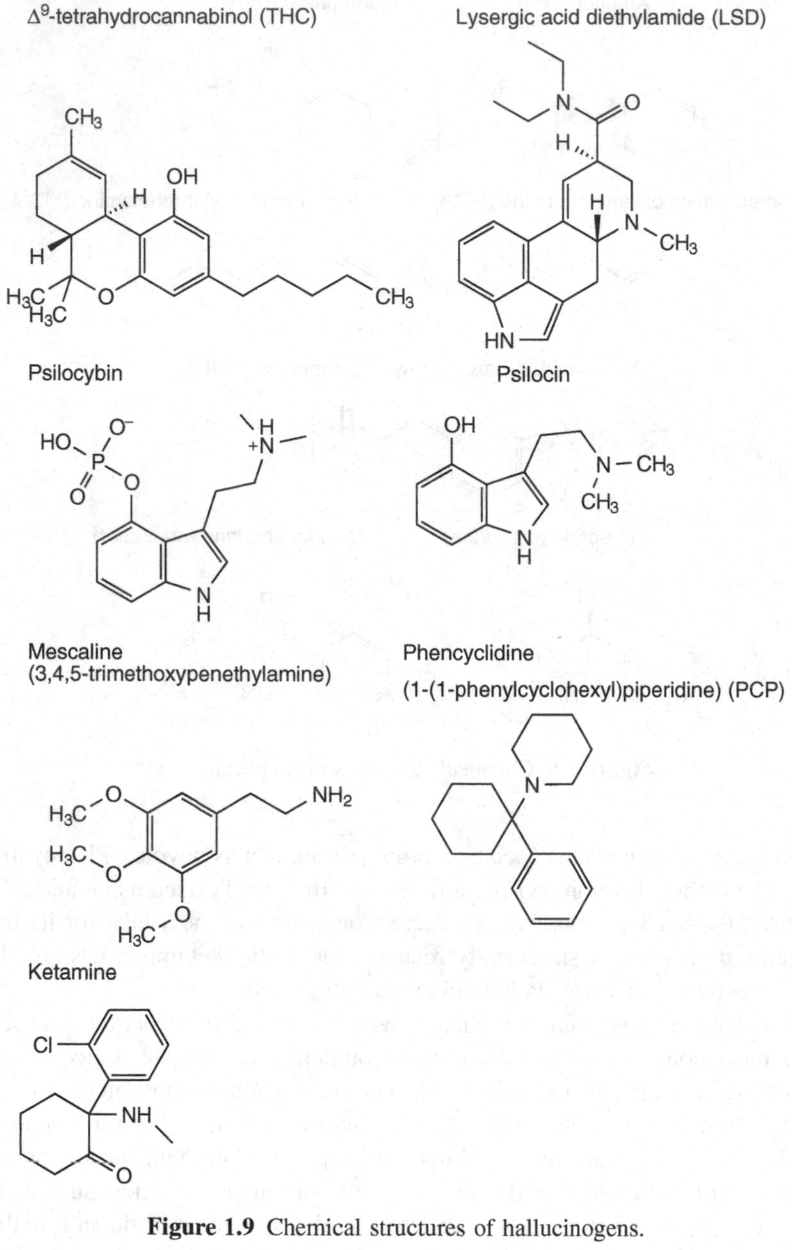

Δ⁹-tetrahydrocannabinol (THC)   Lysergic acid diethylamide (LSD)

Psilocybin   Psilocin

Mescaline
(3,4,5-trimethoxypenethylamine)

Phencyclidine
(1-(1-phenylcyclohexyl)piperidine) (PCP)

Ketamine

Figure 1.9 Chemical structures of hallucinogens.

and water. The remainder of the alcohol is excreted in breath, urine and sweat. The amount of alcohol in breath is directly proportional to the blood alcohol concentration (BAC).

Inhalants are volatile chemicals that are inhaled to produce the desired effect [28, 39]. Examples of substances used as inhalants include paint thinners and petrol. Inhalants produce effects similar to those of alcohol. Some common substances found in inhalants are toluene, butane and halogenated hydrocarbons.

A number of synthetic *CNS depressants* are encountered in forensic samples and may be diverted from legitimate sources [28–30, 33]. Barbiturates have been in use for an extended period and are prescribed for a variety of purposes including as tranquilizers, muscle relaxants or hypnotics. They may be in the form of powders, tablets, capsules or solutions. The widely prescribed benzodiazepines were introduced to replace barbiturates and are commonly found in tablet or capsule form. Gamma-hydroxybutyric acid (GHB) is a hydroxylated short-chain fatty acid and can act as a CNS depressant [28, 33]. The salt can be in the form of powders, tablets or solution. Figure 1.10 shows the structures of some common CNS-depressant drugs.

Carbon monoxide (CO) is an odourless, colourless and tasteless gas produced from the partial combustion of organic compounds and can be produced by, for example, fires and the exhaust from petrol engines [28, 40]. The gas can be inhaled in the lungs and absorbed into the blood. The toxicity of CO results from its combination with haemoglobin (Hb) in red blood cells, coordinating with the iron atoms of Hb to produce carboxyhaemoglobin (COHb). Cyanide is also a toxic material that can be produced in fires [28, 40]. Hydrogen cyanide (HCN) can result from the hydrolysis of nitrogen-containing materials such as polyurethanes and polyacrylonitrile. Cyanide is inhaled and transported to blood. The toxicity of cyanide is a result of its ability to affect the cellular processes of cytochrome.

Poisoning due to particular metals can occur via environmental exposure or ingestion [28]. A number of metals can produce detrimental effects when they reach a high enough concentration in the human body. The determination of metal concentrations in biological samples can provide information about a source of poisoning. Metals that are considered in forensic toxicology include aluminium, arsenic, cadmium, iron, mercury, lithium, lead and thallium. The analysis of hair for heavy metals, such as lead, arsenic and mercury, is well established.

1.2.12 Fingerprints

The detection of fingerprints, or fingermarks, on an object can provide evidence that the object was touched by an individual [41, 42]. Fingermarks are composed of natural secretions and contaminants from the environment. Such impressions are generally invisible, are described as latent and require some form of treatment to make them visible. Hands contain eccrine glands and a deposit produced

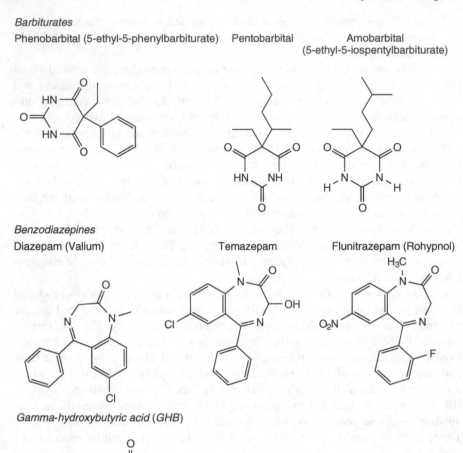

Barbiturates

Phenobarbital (5-ethyl-5-phenylbarbiturate)   Pentobarbital   Amobarbital
                                                              (5-ethyl-5-iospentylbarbiturate)

Benzodiazepines

Diazepam (Valium)   Temazepam   Flunitrazepam (Rohypnol)

Gamma-hydroxybutyric acid (GHB)

Figure 1.10 Chemical structures of CNS depressants.

by these glands is amino acids. Sebaceous glands (located e.g. on the forehead) also contribute lipid materials, such as fatty acids, due to the contamination of fingers by the touching of the face. Fingermarks can be detected on different surface types. Porous surfaces, such as paper, enable latent fingermarks to be rapidly absorbed onto the substrate. Latent fingermarks are not absorbed into non-porous substrates, such as glass or plastic. Such marks are more easily destroyed. Semi-porous surfaces, such as latex gloves and plastic banknotes, have intermediate properties. Once a fingerprint has been suitably enhanced (as described in Chapter 2), it may be classified based on pattern similarities. Automated fingerprint identification can be carried out to match fingerprints.

1.3 Introduction to Data Analysis

A broad range of experimental techniques are used to collect qualitative and quantitative information about the variety of forensic samples, and the subsequent chapters of this book will detail the methods commonly used. Once data are generated, there is often a need to apply a statistical approach to the data produced. There are two roles for statistics in forensic science. Forensic scientists require statistics to interpret the data obtained from experiments, and this data analysis is the focus in this book. Statistics is also required for the next stage, evidence evaluation – that is, the interpretation of observations from casework. For instance, likelihood ratios can be determined to predict a link between a piece of evidence and a crime scene.

A number of statistical approaches are used in the analysis of the forensic data generated by the techniques discussed in this book. A detailed examination of the origins of these statistical methods is beyond the scope of this book, but a number of useful texts are available and several focus on the application of such methods to forensic problems [43–45]. A starting point is the normal distribution of data, with a mean and standard deviation being used to predict the probabilities of particular measurements. Where the normal distribution is not applicable, a t-test can be utilized for the statistical comparison of data.

Linear regression, which involves finding a relationship between two parameters, is also a valuable tool for the quantitative analysis of a variety of forensic data where there is a single dependent variable. Regression lines can be used for calibration purposes and can be produced using an independent variable, x, and a dependent variable, y. The equation used is:

$$y = mx + b \tag{1.1}$$

where m is the slope and b is the y-intercept. Least squares minimization involves fitting data to a line so that there is a minimum deviation from all data points. The quality of the fit of the line is measured by the squared correlation coefficient, R^2.

Classification techniques are also very useful for deciphering complex forensic data. Several *multivariate statistical methods* are available that may be used to quantify or classify data, and these are commonly known as chemometric techniques. A widely used method is principal component analysis (PCA), which is an unsupervised method, meaning that no assumptions are made about the data in advance and individual samples are grouped based on the similarity amongst the data. Hierarchical clustering analysis (HCA) is also an unsupervised technique and clusters data in the form of a dendrogram (a tree-like structure) revealing the similarity or dissimilarity of data. There are also supervised methods that utilize prior assumptions about the existence of groups of data and enable a large number of variables to be compared within a data set. Linear discriminant analysis (LDA) and artificial neural networks (ANNs) are examples of supervised methods.

1.4 Summary

An array of material types are encountered as forensic evidence. Such evidence is used to provide a link to a crime scene or to determine how a crime was committed. An understanding of the chemistry of evidence is fundamental to solving such problems – analysis of evidence requires knowledge of the underlying chemical properties. The following chapters deal with the tools used to analyse the chemical properties of forensic evidence.

References

1. Stuart, B. H., *Polymer Analysis*, John Wiley & Sons Ltd, Chichester, UK, 2002.
2. Watson, N., Forensic sciences/fibres, in *Encyclopedia of Analytical Science*, 2nd ed., P. Worsfold, A. Townshend and C. Poole (Eds), Elsevier, Amsterdam, 2005, pp. 406–414.
3. Eyring, M. B. and Gaudette, B. D. An introduction to the forensic aspects of textile fibre examination, in *Forensic Science Handbook Vol.* **2**, 2nd ed., R. Saferstein (Ed.), Prentice Hall, Upper Saddle River, NJ, 2005, pp. 245–261.
4. Roux, C. and Robertson, J., *Fibres/types*, in Encyclopedia of Forensic Sciences, Academic Press, London, 2000, pp. 838–854.
5. Robertson, J. and Grieve, M. (*Eds*), *Forensic Examination of Fibres*, Taylor and Francis, London, 1999, pp. 160–176.
6. Goodpaster, J. V. and Liszewski, E. A., *Anal. Bioanal. Chem.* **394**, 2009–2018 (2009).
7. *Colour Index*, 4th online ed., Society of Dyers and Colourists and American Association of Textile Chemists and Colourists, London, 2002.
8. Thornton, J. I., Forensic paint examination, in *Forensic Science Handbook Vol.* **1**, 2nd ed., R. Saferstein (Ed.), Prentice Hall, Upper Saddle River, NJ, 2002, pp. 429–478.
9. Bentley, J., Composition, manufacture and use of paint, in *Forensic Examination of Glass and Paint*, B. Caddy (Ed.), Ellis Horwood, London, 2001, pp. 123–142.
10. Buzzini, P. and Stoecklein, W., Forensic sciences/paints, varnishes and lacquers, in *Encyclopedia of Analytical Science*, P. Worsfield, A. Townshend and C. F. Poole (Eds.), Elsevier, Amsterdam, 2005, pp. 453–464.
11. Brun-Conti, L., Paints and coatings, in *Encyclopedia of Forensic Science*, J. A. Siegel (Ed.), Academic Press, London, 2000, pp. 1141–1148.
12. Brunelle, R. L., Document analysis/ink analysis, in *Encyclopedia of Forensic Sciences*, J. A. Siegel (Ed.), Academic Press, London, 2000, pp. 591–597.
13. Brunelle, R. L., Questioned document examination, in *Forensic Science Handbook Vol.* **1**, 2nd ed., R. Saferstein (Ed.), Prentice Hall, Upper Saddle River, NJ, 2002, pp. 697–744.
14. Neumann, C. and Mazzella, W. D., Forensic sciences/questioned documents, in *Encyclopedia of Analytical Science*, 2nd ed., P. Worsfold, A. Townshend and C. Poole (Eds), Elsevier, Amsterdam, 2005, pp. 465–471.
15. Lambert, J. A., Forensic sciences/glass, in *Encyclopedia of Analytical Science*, 2nd ed., P. Worsfold, A. Townshend and C. Poole (Eds.), Elsevier, Amsterdam, 2005.
16. Koons, R. D., Buscaglia, J., Bottrell, M. and Miller, E. T., Forensic glass comparisons, in *Forensic Science Handbook Vol.* **1**, 2nd ed., R. Saferstein (Ed.), Prentice Hall, Upper Saddle River, NJ, 2002, pp. 161–213.
17. Murray, R. C. and Solebello, L. P., Forensic examination of soil, in *Forensic Science Handbook Vol.* **1**, 2nd ed., R. Saferstein (Ed.), Prentice Hall, Upper Saddle River, NJ, 2002, pp. 615–633.
18. Rowe, W. F., Forensic applications, in *Encyclopedia of Soils in the Environment*, D. Hillel (Ed.), Elsevier, Amsterdam, 2005, pp. 67–72.

19. Dawson, L. A., Campbell, C. D., Hillier, S. and Brewer, M. J., Methods of characterising and fingerprinting soils for forensic application, in *Soil Analysis in Forensic Taphonomy*, M. Tibbett and D. O. Carter (Eds), CRC Press, Boca Raton, 2008, pp. 271–315.

20. Sleeman, R. and Carter, J. F., Forensic sciences/explosives, in *Encyclopedia of Analytical Science*, 2nd ed., P. Worsfold, A. Townshend and C. Poole (Eds.), Elsevier, Amsterdam, 2005, pp. 400–406.

21. Midkiff, C. R., Arson and explosive investigation, in *Forensic Science Handbook Vol.* 1, 2nd ed., R. Saferstein (Ed.), Prentice Hall, Upper Saddle River, NJ, 2002, pp. 479–524.

22. Monts, D. L., Singh, J. P. and Boudreaux, G. M., Laser- and optical-based techniques for the detection of explosives, in *Encyclopedia of Analytical Chemistry*, R. A. Meyers (Ed.), John Wiley & Sons, Inc., Hoboken, NJ, 2006.

23. Rowe, W. F., Firearms identification, in *Forensic Science Handbook Vol.* 2, 2nd ed., R. Saferstein (Ed.), Prentice Hall, Upper Saddle River, NJ, 2002, pp. 393–461.

24. Lewis, S. W., Agg, K. M., Gutowski, S. J. and Ross, P., Forensic sciences/gunshot residues, in *Encyclopedia of Analytical Science*, 2nd ed., P. Worsfold, A. Townshend and C. Poole (Eds), Elsevier, Amsterdam, 2005, pp. 430–436.

25. Baron, M., Forensic sciences/arson residues, in *Encyclopedia of Analytical Science*, 2nd ed., P. Worsfold, A. Townshend and C. Poole (Eds.), Elsevier, Amsterdam, 2005, pp. 365–372.

26. Gunn, A., *Essential Forensic Biology*, 2nd ed., John Wiley & Sons Ltd, Chichester, UK, 2009.

27. Brandt-Casadevall, C. and Dimo-Simonin, N., Forensic sciences/blood analysis, in *Encyclopedia of Analytical Science*, 2nd ed., P. Worsfold, A. Townshend and C. Poole (Eds), Elsevier, Amsterdam, 2005, pp. 373–378.

28. Levine, B. (Ed.), *Principles of Forensic Toxicology*, 2nd ed., AACC Press, Washington, DC, 2006.

29. Cole, M. D., *The Analysis of Controlled Substances*, John Wiley & Sons Ltd, Chichester, UK, 2003.

30. Moffat, A. C., Osselton, M. D. and Widdop, B., *Clarke's Analysis of Drugs and Poisons*, 4th ed., Pharmaceutical Press, London, 2011.

31. Drummer, O. H., Toxicology methods of analysis – ante-mortem, in *Encyclopedia of Forensic Sciences*, Academic Press, London, 2000, pp. 1397–1403.

32. Drummer, O. H., Toxicology/methods of analysis – post-mortem, in *Encyclopedia of Forensic Sciences*, Academic Press, London, 2000, pp. 1404–1409.

33. Daèid, N. N., Forensic sciences/illicit drugs, in *Encyclopedia of Analytical Science*, 2nd ed., P. Worsfold, A. Townshend and C. Poole (Eds), Elsevier, Amsterdam, 2005, pp. 446–453.

34. Merves, M. L. and Goldberger, B. A., Heroin, in *Encyclopedia of Analytical Science*, 2nd ed., P. Worsfold, A. Townshend and C. Poole (Eds), Elsevier, Amsterdam, 2005, pp. 260–266.

35. Cody, J. T., Amphetamines, in *Encyclopedia of Analytical Science*, 2nd ed., P. Worsfold, A. Townshend and C. Poole (Eds), Elsevier, Amsterdam, 2005, pp. 80–88.

36. Jones, A. W. (Ed.), Alcohol, in *Drug Abuse Handbook*, 2nd ed., Taylor and Francis, London, 2007.

37. Moriya, F., Forensic sciences/alcohol in body fluids, in *Encyclopedia of Analytical Science*, 2nd ed., P. Worsfold, A. Townshend and C. Poole (Eds), Elsevier, Amsterdam, 2005, pp. 358–365.

38. Emerson, V. J., Alcohol analysis, in *Crime Scene to Court: Essentials of Forensic Science*, 2nd ed., P. White (Ed.), *Royal Society of Chemistry*, Cambridge, 2004, pp. 350–376.

39. Flanagan, R. J., Forensic sciences/volatile substances, in *Encyclopedia of Analytical Science*, 2nd ed., P. Worsfold, A. Townshend and C. Poole (Eds), Elsevier, Amsterdam, 2005, pp. 486–498.

40. Gray, C. N. Forensic sciences/carbon monoxide and cyanide from fire and accident, in *Encyclopedia of Analytical Science*, 2nd ed., P. Worsfold, A. Townshend and C. Poole (Eds), Elsevier, Amsterdam, 2005, pp. 379–383.

41. Lennard, C., Forensic science/fingerprint techniques, in *Encyclopedia of Analytical Science*, 2nd ed., P. Worsfold, A. Townshend and C. Poole (Eds), Elsevier, Amsterdam, 2005, pp. 414–423.

42. Lennard, C., *Platypus* **82**, 23–27 (2004).

43. Lucy, D., *Introduction to Statistics for Forensic Scientists*, John Wiley & Sons Ltd, Chichester, UK, 2005.
44. Adam, C., *Essential Mathematics and Statistics for Forensic Science*, John Wiley & Sons Ltd, Chichester, UK, 2010.
45. Zadora, G., Chemometrics and statistical considerations in forensic science, in *Encyclopedia of Analytical Chemistry*, R. A. Meyers (Ed.), John Wiley & Sons Ltd, Chichester, UK, 2010.

Chapter 2
Preliminary Tests

Learning Objectives

- To apply simple chemical tests to the identification of forensic samples including drugs, toxicological specimens, body fluids, gunshot residues, paint and questioned documents.
- To use density measurements to identify forensic samples including glass, polymer and soil samples.
- To understand how different light sources combined with chemical treatments can be used to obtain information about forensic samples including fingerprints, body fluids and documents.

2.1 Introduction

For particular types of forensic evidence, the first stage is to carry out simple test procedures to detect the presence of a suspected material. Often such presumptive tests are based on chemical reactions using small quantities of the sample of interest. The results of such tests enable an appropriate choice of confirmatory test to be made. Confirmatory tests provide more specific information, and such tests will be described in subsequent chapters. Another initial step is to make a visual record of evidence. There is a range of techniques using different light sources that enhance the visual information to be obtained for forensic evidence.

2.2 Chemical Tests

There are several simple chemical test types available that enable particular substances to be quickly identified. Colour and solubility tests can be carried out

Forensic Analytical Techniques, First Edition. Barbara Stuart.
© 2013 John Wiley & Sons, Ltd. Published 2013 by John Wiley & Sons, Ltd.

using simple equipment [1–4]. These approaches are destructive, but if enough sample is available, they provide a rapid straightforward approach before more complex techniques are utilized.

2.2.1 Methods

Colour (or *spot) tests* involve the observation of a colour change when a specific reagent is reacted with a sample. Generally a small quantity of sample is placed on a spot plate or in a test tube and the reagent added. Positive controls, where the substance to be detected is known to be present, and negative controls (or blanks) need to be carried out at the same time. This eliminates the possibility of false positives due to, for instance, the presence of contaminants or degradation products. *Solubility tests* can also provide a useful screening method. The requirement of the examination of small specimens means that a small quantity of solvent should be dropped on the specimen and observations made using a stereomicroscope (described in Chapter 3).

2.2.2 Drugs and Toxicology

A number of simple preliminary tests are available for the screening of drugs [1, 2, 5–8]. Examples of some of the more common colour tests are listed here:

- *Marquis test*: 1 drop of 5 ml 40 v/v% formaldehyde in 100 ml concentrated H_2SO_4 is directly added to the sample. A violet colour indicates that codeine, heroin, morphine or opium is present. An orange-brown colour indicates an amphetamine or methamphetamine.

- *Mandelin test*: 0.5–1.0 w/v% ammonium vanadate in H_2SO_4 is directly added to the sample. A grey colour indicates LSD. A green-blue colour indicates methadone or MDA. A blue-grey colour indicates morphine.

- *Ehrlich test*: 1 w/v% p-dimethylaminobenzaldehyde in 10% HCl in ethanol is directly added to the sample. A violet colour indicates LSD. A blue-grey colour indicates psilocin. A red-brown colour indicates psilocybin.

- *Scott test*: A drop of 2 w/v% cobalt thiocyanate in water and glycerine (1:1 volume ratio) is added to the sample. Cocaine will produce a blue colour. The addition of 1 drop of concentrated HCl changes a blue colour to pink. Further confirmation is provided by the addition of several drops of chloroform, which shows a blue colour.

- *Duquenois–Levine test*: 1 drop of ethanol is added to the sample followed by 5 drops of 2 g vanillin and 2.5 ml acetaldehyde in 100 ml of 95 v/v% ethanol. 10 drops of chloroform are then added after shaking. A blue-purple colour indicates cannabis.

- *Dille–Koppani test*: Several drops of 1 w/v% cobalt nitrate in methanol followed by several drops of 5 v/v% isopropylamine in methanol are added to the sample. A blue colour indicates the presence of barbiturates.

SAQ 2.1

Several standard colour tests were carried out on a suspect sample believed to contain a drug. The Ehrlich test produced a violet colour, and the Mandelin test produced a grey colour in the sample. Is it possible to identify what drug type is present in this sample based on these test results?

Immunoassays, tests in which antibodies are used, are also employed to detect the presence of drugs or poisons in body fluids such as urine or blood. Immunoassays are based on antibody–antigen reactions: a specific antibody is produced by the immune system in response to the presence of a particular substance. The antibody selectively binds to the antigen to form an antigen–antibody complex. The tests are designed to detect a specific analyte. Although immunoassays are fast and sensitive, users must be aware that the presence of adulterants can cause false negative results.

2.2.3 Body Fluids

There are a number of established catalytic colour tests available for the presumptive testing of bloodstains [4, 9]. The tests are based on the fact that haemoglobin behaves as an enzyme in the presence of hydrogen peroxide, which can catalyse oxidation reactions and produce a colour change. For such tests, the sample of interest is collected with a cotton swab and a drop of reagent is added followed by a drop of hydrogen peroxide. The common tests employed are summarized in Table 2.1. Although these tests are sensitive, they do have the disadvantage of not being specific for blood. The tetramethylbenzidine (TMB) and o-tolidine tests are more sensitive methods, but the phenolphthalein (Kastle–Meyer) and leucomalachite green tests are more selective. There are also simple immunological confirmation tests for blood available that are based on the detection of haemoglobin [10]. They are based on the interaction of haemoglobin antibodies

Table 2.1 Colour tests for bloodstains

Test reagent	Positive result colour
phenolphthalein (Kastle–Meyer)	pink
o-tolidine	blue
leucomalachite green	green
tetramethybenzidine (TMB)	blue-green

binding to blue dye molecules. The presence of the blue antigen–antibody complex indicates a positive result, and the test is sensitive.

SAQ 2.2

A possible bloodstain is examined using both the TMB and Kastle–Meyer tests. The TMB test produces a positive result, while the Kastle–Meyer test is negative. What is a possible reason for these apparently contradictory test results?

A common preliminary test for detecting the presence of semen involves identifying the enzyme *seminal acid phosphatase* (SAP) [4, 9]. The enzyme catalyses the hydrolysis reactions of organic phosphates and forms a product that reacts with a diazonium salt chromagen to produce a colour change. Popular reagent components are α-naphthyl phosphate and Brentamine Fast Blue. A cotton swab or filter paper is used to collect a sample from a stain. The reagent is applied, and a purple colour indicates the presence of a seminal stain.

A common confirmatory test for semen is known as the Christmas tree stain and is used to establish the presence of sperm. The test involves the addition of red and green dyes to an unknown stain that has been collected and smeared on a glass slide: staining distinguishes between the sperm cells from the epithelial cells. The *prostate-specific antigen* (PSA) test can also be used to test for the presence of semen in the absence of sperm. Commercially available test kits are available and rely on the interaction between the PSA produced by the prostate gland and an antibody.

The most common chemical test for saliva is the *starch–iodine test*, which is based on the detection of the enzyme amylase [4, 9]. Starch appears blue in the presence of iodine, and amylase causes the starch to break down and the blue colour disappears. Thus, the absence of a blue colour is a positive test for saliva. The presence of proteins, such as albumin and γ-globulin, can produce false positives. The Phadebas® reagent, in which starch is linked to a dye molecule to form an insoluble complex, is a common commercial product. As the starch breaks down, the dye becomes soluble and a quantitative colour change is observed. A tube or a press test, where the reagent is applied to filter paper which is then applied to the region of interest, are common testing methods.

2.2.4 Gunshot Residue

There are three common chemical tests used to identify the presence of GSR: the modified Griess test, the dithiooxamide test and the sodium rhodizonate test [4, 8, 11]. The *modified Griess test* is used to detect the presence of nitrite compounds. The evidence under investigation, such as an item of clothing, is placed face down on desensitized photographic paper that has been treated with H_2SO_4 in water and α-naphthol in methanol. The item is heated with acetic acid solution to form nitrous acid. The nitrous acid reacts with the sulfanilic acid to form a diazonium compound, which reacts with the α-naphthol to form an orange azo dye. The paper can then be used to determine the distribution of GSR and, hence, determine the firing distance.

The *dithiooxamide test* (also known as the rebeanic acid test) is used to detect the presence of particular metal elements. The material of interest is exposed to an ammonia solution followed by exposure to a diothiooxamide ethanol solution. The development of a green-grey colour indicates the presence of copper, while the presence of nickel is indicated by a blue-pink colour.

The *sodium rhodizonate test* detects the presence of lead. The test area is sprayed with an aqueous sodium rhodizonate solution followed by a sodium bitartrate–tartaric acid buffer solution. The appearance of a pink colour indicates the presence of a heavy metal. The presence of lead can be confirmed by spraying the test area with dilute HCl. The pink colour becomes a violet-blue colour if lead is present.

The modified Griess test should be carried out first as it does not interfere with the results of the dithiooxamide test or the sodium rhodizonate test. The sodium rhodizonate test can also potentially interfere with nitrite testing.

SAQ 2.3

The dithiooxamide test is carried out on GSR on an item of white clothing, and a green-grey colour results. What is the element identified by this test? What is the source of this element?

2.2.5 Explosives

A number of recognized spot tests can be used to detect the presence of common explosives or explosive residue [6, 11]. Explosive materials are often extracted into acetone. A Griess test can be used to detect nitrate ions, and a simple silver nitrate precipitation test can be used to identify the presence of chloride ions. The presence of ammonium ions can be identified using Nessler reagent (HgI_2, KI and KOH in water): a drop of reagent forms an orange-brown precipitate if ammonium ions are present.

2.2.6 Paint

Solubility tests can be carried out on paint in order to discriminate specimens that cannot be separated based on appearance alone. The technique is based on the principle that different binder compositions show different solubilities [3, 8, 12]. Some common solvents include acetone, chloroform and dichloromethane. A variety of microchemical tests are also available that allow colour changes associated with pigment and binder reactions to discriminate paint specimens [13]. There is a range of tests based on dehydration, oxidation or reduction reactions that produce a colour change in organic molecules. Two commonly used tests are the use of diphenylamine in concentrated H_2SO_4 and glacial acetic acid and the Le Rosen test, which involves a solution of formaldehyde in concentrated H_2SO_4.

2.2.7 Documents

Chemical tests can be used to characterize the ink type used in questioned documents [14, 15]. Solubility tests provide a preliminary means of determining whether samples are composed of the same ink. Small amounts of solvents can be directly applied to ink, or an ink sample may be removed for testing. Generally, ballpoint and marker pens are soluble in pyridine, and inks for roller pens and fountain pens are soluble in ethanol or water.

2.3 Density

The determination of density provides a simple preliminary means of identification for a range of materials [16, 17]. Density (ρ) is the ratio of the mass (m) of a substance to its volume (V):

$$\rho = \frac{m}{V} \tag{2.1}$$

Density measurements are made by exploiting Archimedes' principle, which states that an object immersed in a fluid is buoyed by a force equal to the weight of a displaced fluid.

2.3.1 Methods

There are several qualitative and quantitative approaches to density measurements for forensic specimens. The *flotation method* utilizes the phenomena that a solid material will float in a liquid of greater density, sink in a liquid of lower density and remain suspended in a liquid of equal density.

Another technique is the *density gradient method*, where a tube containing liquids mixed in varying proportions is employed. The density of the liquid for each layer is determined using:

$$\rho = \frac{\rho_1 V_1 + \rho_2 V_2}{V_1 + V_2} \tag{2.2}$$

where ρ is the density of the mixture, ρ_1 is the density of the first liquid component, ρ_2 is the density of the second liquid component, V_1 is the volume of the first liquid and V_2 is the volume of the second liquid. The gradient tube will have a series of layers of varying density and when a solid specimen is added to the tube, it will sink until it reaches a level where its density is the same as the liquid at that height. The gradient tube can be calibrated by adding solids of known density, and an estimation of the absolute density of an unknown can be made.

SAQ 2.4

A stock solution of sodium polytungstate in aqueous solution has a density of $2.89\,g\,cm^{-3}$ at $25°C$. What will be the density of a 50 v% sodium polytungstate and 50 v% water solution to be used for a density gradient experiment?

Density measurements can also be made using a modified mass balance, which can consist of a hanging support with a hook that rests on the balance pan, a platform above the balance pan and a metal basket with hanging wires that can be submerged into a beaker. Measurements are made in air and on immersion in a liquid at a constant temperature. The density of the specimen is measured using the equation:

$$\rho = \frac{w_1 \times \rho_{liquid}}{w_1 - w_2} \tag{2.3}$$

where w_1 is the weight of the specimen in air, w_2 is the weight of the specimen immersed in the chosen liquid and ρ_{liquid} is the density of the liquid in which the specimen is immersed.

As density is a temperature-dependent quantity, it is important to record the temperature at which density measurements are carried out.

2.3.2 Glass

Density measurements provide a simple means of determining if glass speci-mens originate from the same source [8, 15, 16, 18]. One approach is to use a density gradient method employing the liquids bromoform ($\rho = 2.89\,g\,cm^{-3}$) and bromobenzene ($\rho = 1.52\,g\,cm^{-3}$). Glass specimens should be clean and dry before density measurements are carried out. Glasses have densities of the order of $2\,g\,cm^{-3}$ and the exact value depends on the chemical structure of the glass.

SAQ 2.5

A series of bromoform–bromobenzene solutions is prepared for use in a density gradient experiment and the details are listed in Table 2.2. A glass specimen is immersed in the solutions and the observations made are also provided in Table 2.2. Estimate the density of the glass specimen.

2.3.3 Soil

The density of the mineral content of soil samples can provide a useful prelimi-nary comparison of an unknown soil specimen with that collected from a crime scene [8, 16, 20]. The density gradient method is usually employed and sodium polytungstate–water or bromoform–bromobenzene solutions are common choices

Table 2.2 Bromoform–bromobenzene solutions and observations for density gradient measurements (SAQ 2.5)

Bromoform (%)	Bromobenzene (%)	Density (g cm^{-3})	Observation
100	0	2.90	Floats
80	20	2.62	Floats
60	40	2.34	Floats
40	60	2.06	Sinks
20	80	1.78	Sinks
0	100	1.50	Sinks

for the procedure. Prior to analysis, the soil samples are passed through a series of wire sieves with a range of mesh sizes, and the 40 to 60 mesh fractions are usually used for density experiments.

DQ 2.1

Could the density of the mineral content of a soil sample determined using a density method on one day be compared to the value determined the next day in the same laboratory?

Answer

As density is a temperature-dependent quantity, the temperature at which the experiment is carried out needs to be considered in order for a valid comparison to be made. If the laboratory temperature changes, then different densities will be recorded for reference solutions and specimens.

2.3.4 Polymers

Density measurements are an established means of discriminating polymer specimens, and quite well-defined density ranges can be determined for pure polymers [19]. Polymers have densities of the order of 1 g cm^{-3}. If access to an appropriate balance is unavailable, a flotation approach can be taken using a variety of liquids of known density. For example, specimens can be immersed in methanol ($\rho = 0.79$ g cm^{-3}), water ($\rho = 1.00$ g cm^{-3}), saturated NaCl solution ($\rho = 1.20$ g cm^{-3}) and saturated MgCl$_2$ solution ($\rho = 1.34$ g cm^{-3}). It is noted that the presence of additives or different processing methods can result in variation of polymer density values. For example, a foam polymer will have a lower density than a moulded version of the same polymer type.

SAQ 2.6

Car headlamp fragments are collected, and the density is to be used to confirm whether the fragments are made of glass or plastic. The fragments are placed in a saturated NaCl solution. What information is gained if the fragments are observed to sink in this experiment? What further test could be carried out to provide an estimation of the density of the fragment?

2.4 Light Examination

A common starting point for the investigation of evidence is optical examination using normal light, often with magnification. Subsequent light examination techniques can be used, including reflection and luminescence techniques using visible, ultraviolet and infrared radiation. Photography is an important means of documenting forensic evidence, and traditional photography records information using visible light, but other parts of the electromagnetic spectrum can also be used to produce images. Ultraviolet (UV) and infrared radiation can be used to enhance the information provided by forensic evidence. For instance, infrared reflected photography can be used to detect substances that are not visible by eye.

A useful phenomenon that can result from exposure to a particular wavelength is *photoluminescence* [21]. One of the most widespread applications of luminescence in forensic science is for the enhancement of images. When molecules absorb radiation in electronic transitions to form excited states, the energy can undergo several relaxation mechanisms. If energy loss occurs via the emission of radiation, a process known as photoluminescence occurs. *Fluorescence* is a form of photoluminescence that involves emission occurring from the lowest excited singlet electronic state to the singlet ground electronic state. *Phosphorescence* is another form of photoluminescence and involves the emission that occurs from the lowest excited triplet electronic state to the singlet ground electronic state. Luminescence can also result from a chemical reaction and is not dependent on the light source – this is known as *chemiluminescence*.

2.4.1 Methods

Alternate light sources (ALSs) have been designed specifically for forensic applications. A forensic light source is composed of a powerful light source, usually a xenon arc lamp. ALSs are tunable to light in the ultraviolet, visible and infrared regions of light. Reflectance, absorption and fluorescence experiments can be carried out using such devices.

2.4.2 Fingerprints

Various light sources combined with surface treatments are used to detect latent fingermarks [21, 22]. Some latent fingermarks exhibit an inherent luminescence, which is produced by excitation with blue-green light produced by an argon laser. The luminescence is believed to result from the presence of contaminants. Inherent luminescence is useful as it is nondestructive, but as it is weak it can be used only when the background fluorescence of the substrate is minimal.

A common approach to fingermark detection on nonporous surfaces is the use of powder. Fingerprint *powders*, such as carbon black or graphite, can be applied to a surface using a brush. The powder adheres to the fingermark deposit and the fingermark can be lifted using adhesive tape or film. The enhanced fingerprint is observed and photographed. Powdering works best for freshly produced marks. *Staining* a fingerprint with a fluorescent dye is another approach for nonporous surfaces. A common approach is cyanoacrylate *fuming*. Cyanoacrylate esters selectively polymerize to produce a white solid, polycyanoacrylate, on latent fingermark ridges at ambient temperature in an enclosed chamber. The technique can be enhanced by the application of a luminescent dye, such as rhodamine 6G. *Vacuum metal deposition* (VMD) can also follow cyanoacrylate fuming when background interference exists. A thin layer of gold is deposited on the surface to reveal ridges. Fingerprint detail can be obtained for surfaces such as glass, plastic or metal. The recommended sequence for fingerprint detection on nonporous surfaces is: optical → cyanoacrylate fuming → VMD → luminescent dye.

The examination of fingermarks on porous surfaces, such as paper, requires chemical modification of a fingermark component to produce a photoluminescent compound. A widely used reagent for porous surfaces is *ninhydrin*, which reacts with the amino acids of a fingermark to produce a purple compound known as Ruhemann's purple (the reaction is illustrated in Figure 2.1). The reagent is initially dissolved in a small amount of a polar solvent and acetic acid, and then made up in a nonpolar solvent. The reagent is applied using a spray or a brush, and the fingerprint dried. The purple images are usually photographed under

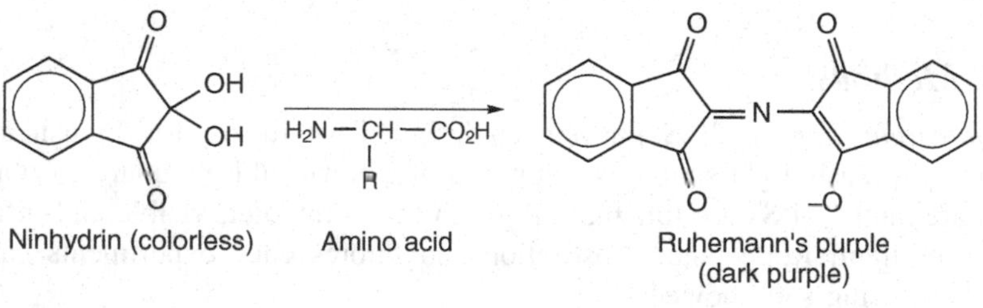

Ninhydrin (colorless)   Amino acid   Ruhemann's purple
                                                          (dark purple)

Figure 2.1 Reaction of ninhydrin with amino acids. Reproduced with permission from the Journal of Forensic Sciences, Vol 32, 597, 1987. Copyright ASTM International, 100 Barr Harbor Drive, West Conshohocken, PA 19428.

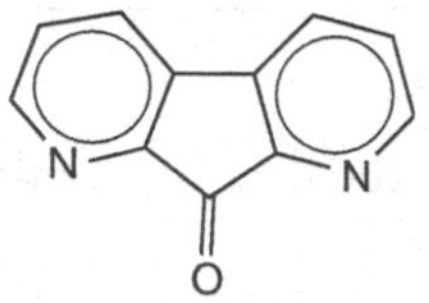

Figure 2.2 Chemical structure of 1,8-diafluoren-9-one.

Figure 2.3 Fingerprint developed with DFO.

white light. Improvements to a ninhydrin fingerprint can be made by modification with a metal salt solution, such as zinc chloride. A complex formed between Ruhemann's purple and the metal ion produces a luminescence that enhances the fingerprint. Another reagent for developing latent fingermarks on paper is *1,8-diazafluoren-9-one* (DFO) (the structure is shown in Figure 2.2). Heat is required to develop DFO fingerprints, which are fluorescent using excitation at 470–550 nm and producing emission at 570–620 nm. Figure 2.3 illustrates a fingerprint on paper developed with DFO using an excitation of 505 nm. *Physical developer* (PD) is another technique for porous surfaces. The reagent contains silver ions in a Fe^{2+}–Fe^{3+} solution. Silver metal deposits on the print ridges due to the redox reaction: $Ag^+(aq) + Fe^{2+}(aq) \rightarrow Ag(s) + Fe^{3+}(aq)$. As PD is sensitive to water-insoluble fingerprint components, it can be used even if the surface is wet. The recommended sequence for fingerprint detection for porous surfaces is: optical $\rightarrow$ DFO $\rightarrow$ ninhydrin $\rightarrow$ metal salt treatment $\rightarrow$ PD.

SAQ 2.7

Which detection method would you choose for fresh fingermarks on the following substrates:

(a) a polymer banknote.

(b) a paper banknote.

2.4.3 Body fluids

Body fluids, including blood, semen and saliva, can be located using an ALS [9, 23, 24]. Blood produces a strong absorption at 415 nm. The excitation range for semen is broader, and fluorescence can be produced using wavelengths in the range of 300–480 nm. Saliva can also be visualized using ultraviolet light, but is difficult to distinguish from other body fluids.

Bloodstains can also be detected using a reagent sprayed on the suspect bloodstain to enhance its presence [4, 9]. The *luminol test* is an established sensitive presumptive blood test. An aqueous mixture of luminol (3-aminophthalhydrazide) and an oxidant is sprayed over the region of interest, and a blue-white to yellow-green glow indicates the presence of blood. Haemoglobin accelerates the oxidation of luminol in an alkaline solution. Another test for bloodstains involves the use of *fluorescein*. Haemoglobin accelerates the oxidation of fluorescein in hydrogen peroxide. Fluorescein exhibits fluorescence when exposed to ultraviolet light in the range of 425–485 nm.

DQ 2.2

Is the process observed in a luminol test an example of photoluminescence or chemiluminescence?

Answer

As no additional light is required for the luminol reaction to occur, this is an example of a chemiluminescence process. In chemiluminescence, light is emitted as a result of a chemical reaction, while fluorescence is observed when a molecule is exposed to a particular wavelength of light.

2.4.4 Documents

Document examination often begins with nondestructive optical methods [14, 15]. Normal light can be used to examine ink appearance. ALSs can also be used to determine differences in inks that appear similar under normal light conditions and underwriting in documents.

2.5 Summary

When forensic evidence is collected and ready for analysis, several types of preliminary tests can be carried out to decide on the appropriate course of action. A range of simple chemical tests are available to identify evidence such as drugs, toxicological specimens, body fluids, gunshot residues, paint and questioned documents. Density measurements can be carried out to identify forensic samples including glass, polymer and soil samples. Different light sources, often in combination with chemical treatments to the sample, can be used to obtain information about forensic specimens. Light sources, which can include the visible, ultraviolet and infrared spectral regions, are used to identify and characterize fingerprints, body fluids and questioned documents.

References

1. Levine, B., Presumptive chemical tests, in *Encyclopedia of Forensic Sciences*, J. Siegel, G. Knupfer and P. Saukko (Eds), Academic Press, New York, 2000, pp. 167–172.
2. Jungreis, E., *Spot Test Analysis: Clinical, Environmental, Forensic and Geochemical Applications*, John Wiley & Sons, Inc., New York, 1985.
3. Wilson, L. J. and Wheeler, B. P., Optical microscopy in forensic science, in *Encyclopedia of Analytical Chemistry*, R. A. Meyer (Ed.), John Wiley & Sons, Inc., New York, 2009.
4. James, S. and Nordby, J. *(Eds), Forensic Science: An Introduction to Scientific and Investigative Techniques*, CRC Press, Boca Raton, 2003.
5. Siegel, J. A., Forensic identification of controlled substances, in *Forensic Science Handbook Vol. 2*, R. Saferstein (Ed.), Prentice Hall, Englewood Cliffs, NJ, 1988, pp. 68–160.
6. Jungreis, E., Spot tests, in *Encyclopedia of Analytical Science*, 2nd ed., P. Worsfold, A. Townshend and C. Poole (Eds), Elsevier, Amsterdam, 2005, pp. 383–400.
7. Cole, M. D. *The Analysis of Controlled Substances*, John Wiley & Sons, Ltd, Chichester, 2003.
8. Wheeler, B. P. and Wilson, L. J., *Practical Forensic Microscopy: A Laboratory Manual*, John Wiley & Sons, Ltd, Chichester, 2008.
9. Virkler, K. and Ledner, I. K., *Forensic Sci. Int.* **188**, 1–17 (2009).
10. Brandt-Casadevall, C. and Dimo-Simonin, N., Forensic sciences/blood analysis, in *Encyclopedia of Analytical Science*, 2nd ed., P. Worsfold, A. Townshend and C. Poole (Eds), Elsevier, Amsterdam, 2005, pp. 373–378.
11. Feigl, F. and Anger, V., *Spots Tests in Organic Analysis*, 7th ed., Elsevier, Amsterdam, 1989.
12. Thornton, J. H., Forensic paint examination, in *Forensic Science Handbook Vol.* 1, 2nd ed., R. Saferstein (Ed.), Prentice Hall, Englewood Cliffs, NJ, 2002, pp. 429–478.
13. Feigl, F. and Anger, V., *Spot Tests in Inorganic Analysis*, 6th ed., Elsevier, Amsterdam, 1988.
14. ASTM Standard E1422, *Standard Guide for Test Materials for Forensic Writing Ink Comparison*, American Society for Testing and Materials, West Conshohocken, PA, 2005.
15. Ellen, D., *Scientific Examination of Documents: Methods and Techniques*, 3rd ed., CRC Press, Boca Raton, 2005.
16. Meloan, C. E., James, R. E., Brettell, T. and Saferstein, R., *Lab Manual for Criminalistics: An Introduction to Forensic Science*, 10th ed., Prentice Hall, Boston, 2011.
17. Saccocio L. A. and Carroll, M. K., *J. Chem. Ed.* **83**, 1187–1189 (2006).
18. Koons, R. D., Buscaglia, J., Bottrell M. and Miller, E. T., Forensic glass comparisons, in *Forensic Science Handbook Vol.* 1, 2nd ed., R. Saferstein (Ed.), Prentice Hall, Upper Saddle River, NJ, 2002, pp. 161–214.

19. Braun, D., *Simple Methods for Identification of Plastics*, Hanser, Munich, 1996.
20. Chaperlin, K. and Howarth, P. S., *Forensic Sci. Int.* **23**, 161–177 (1983).
21. Menzel, E. R., Fluorescence in forensic science, in *Encyclopedia of Analytical Chemistry*, R. A. Meyers (Ed.), John Wiley & Sons, Inc., New York, 2006.
22. Lennard, C., Forensic sciences/fingerprint techniques, in *Encyclopedia of Analytical Science*, 2nd ed., P. Worsfold, A. Townshend and C. Poole (Eds), Elsevier, Amsterdam, 2005, pp. 414–423.
23. Vandenberg, N. and van Oorschot, R. A. H., The use of Polilight® in the detection of seminal fluid, saliva and bloodstains and comparison with conventional chemical-based screening tests, *J. Forensic Sci.* **51**, 361–370 (2006).
24. Stoilovic, M., *Forensic Sci. Int.* **51**, 289–296 (1991).

Chapter 3

Microscopic Techniques

Learning Objectives

- To understand optical microscopy and how to obtain data from the technique for forensic samples.
- To apply optical microscopy to the study of fibres, paint, drugs, glass, soil, documents and firearms.
- To understand transmission electron microscopy and how to obtain data from the technique for forensic samples.
- To apply transmission electron microscopy to the study of paint.
- To understand scanning electron microscopy and how to obtain data from the technique for forensic samples.
- To apply scanning electron microscopy to the study of gunshot residue, paint, fibres, documents and glass.
- To understand atomic force microscopy and how to obtain data from the technique for forensic samples.
- To apply atomic force microscopy to the study of fibres and documents.
- To understand the origins of X-ray diffraction and how to obtain data from the technique for forensic samples.
- To apply X-ray diffraction to the study of explosives, paint, drugs, documents and soil.

3.1 Introduction

A means of observing properties not readily seen by the naked eye is through the use of optical (or light) microscopy. Optical microscopy has been widely available

Forensic Analytical Techniques, First Edition. Barbara Stuart.
© 2013 John Wiley & Sons, Ltd. Published 2013 by John Wiley & Sons, Ltd.

for many years and enables an object to be magnified many times. Even higher magnifications can be attained using electron microscopy, and valuable structural information can be gained using this approach. The more recent technique of atomic force microscopy also shows promise as a tool in forensic science due to its ability to study specimens at the atomic level. Also included in this chapter is the technique of X-ray diffraction, which enables additional microstructural information to be gathered for particular types of evidence.

3.2 Optical Microscopy

Microscopes are valuable tools for forensic scientists and are used to examine a broad range of evidence types [1–5]. Optical (or light) microscopes employ lenses to magnify objects using the phenomenon of refraction. Structural and chemical information can be enhanced by the use of a microscope. The main types of light microscopes used in forensic laboratories are a simple compound microscope, a comparison microscope, a stereomicroscope, a polarizing light microscope and/or a fluorescence microscope.

3.2.1 Methods

Light microscopes consist of an optical system composed of a light source, a condenser, an objective lens and an ocular lens. Light from the source is collected by a condenser. The focussed light then interacts with the sample on a stage and the image is magnified by the objective. The ocular lens provides a focussed and magnified image for the viewer. A schematic diagram showing a typical optical path for a compound light microscope is shown in Figure 3.1. The degree of magnification will depend on the apparatus design, but up to $\times 1300$ magnifications are achievable. An alignment process, known as a Köhler illumination, is recommended in order to obtain the best image and focus for a specimen, especially as photomicrographs are commonly recorded.

A *comparison microscope* is used to compare forensic specimens side by side. The apparatus consists of a combination of two compound microscopes and allows images from the two microscopes to be viewed in a single field of view. A *stereomicroscope* also consists of two compound microscopes that are aligned to produce a three-dimensional image of a specimen using reflected light. Stereomicroscopy is very useful for the preliminary examination of samples at magnifications ranging from about $\times 2$ to $\times 100$.

A *polarizing light microscope* (PLM) is a valuable microscopic tool in forensic science as it provides additional information about oriented samples such as fibres or minerals. A PLM uses two polarizing filters, a polarizer beneath the stage and an analyser above the objective, oriented perpendicular to each other. The technique can be used to extract information from anisotropic materials, which are materials that have optical properties that vary with the orientation of the

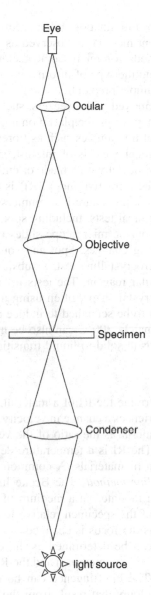

Figure 3.1 Components of a compound light microscope.

incident light. Such materials act as beam splitters, dividing the light rays into two parts. By contrast, isotropic materials show the same optical properties in all directions.

A *fluorescence microscope* can be employed to enhance images where fluorescence is observed in a specimen. A light source with shorter wavelengths, such as UV light, is used for illumination. Filters remove the contribution of the excitation light. A fluorescent material is observed as a bright image.

Sample preparation methods for microscopic techniques vary depending on the nature of the material and the type of information to be obtained. Specimens can be examined with no sample preparation using a stereomicroscope and constituents can be manually removed with tweezers, such as when examining glass or soil samples. A sample may simply be placed on a glass slide with a coverslip. A drop of glycerine is useful for samples such as fibres to hold in place in a wet mount. Where a cross-sectional view is of interest, such as for fibres or paint specimens, the specimen can be cut with a blade or embedded in a resin, then cut with a microtome. Where the refractive index (RI) is to be measured for specimens such as glass or fibres, a procedure is to immerse the specimen in a liquid with a similar RI. Microchemical tests, including spot, solubility and microcrystalline tests, can be carried out on microscope slides or filter paper by adding a drop of solvent and observing changes. Spot and solubility tests were outlined in Chapter 2. During a microcrystalline test, a substance forms a characteristic crystal structure in a particular reagent. The tests involve adding a reagent to a sample and observing the crystal shape, often using a stereomicroscope. Sometimes the mixture may need to be scratched to induce crystal formation. Melting temperatures up to approximately 300°C can also be measured using a hot stage and observing the characteristic solid-to-liquid transition.

3.2.2 Interpretation

A PLM can be used to determine the RI of a material. When light travels into a transparent material, it experiences a change in velocity and is bent at the interface (refraction). The RI of a material is the ratio of the velocity in a vacuum to the velocity in some medium. The RI is a temperature-dependent property that can be used in the identification of materials. A common method of measuring the RI is known as the *Becke line method*. The Becke line is the bright halo near the boundary of a specimen mounted in a medium of different RI. The method involves measuring the RI of the specimen relative to the medium by observing changes to the Becke line as the focus is changed.

The *birefringence* (Δn) can be determined as the difference between the RI parallel to the axis through the specimen ($n_{||}$) and the RI perpendicular to the axis across the specimen ($n_{\perp}$). The birefringence can be used to identify materials based on the interference colours that result from the constructive and destructive interference occurring in an anisotropic substance. The interference colours or retardation of the light is dependent on the thickness and the birefringence. The information has been collected on a Michel–Levy chart, which shows the retardation colour on one axis, the thickness on the other axis and the slopes showing the corresponding birefringence. The ratio of parallel and perpendicular RI values can also be used to determine the sign of elongation for materials such as fibres. The sign of elongation is determined by comparing the values of $n_{||}$

and $n_\perp$. If $n_\parallel$ is larger than $n_\perp$, the sign of elongation is positive, while the sign is negative if $n_\perp$ is larger than $n_\parallel$.

3.2.3 Fibres

Microscopic techniques are regularly employed to identify, characterize and compare synthetic and natural fibres of forensic interest [1–3, 6–8]. A stereomicroscope or comparison microscope can be initially used to examine features such as diameter, surface structure, shape and colour, to quickly determine if fibres belong to the same fibre group.

Natural and synthetic fibres can be discriminated based on physical characteristics viewed microscopically. For example, cotton shows characteristic twists in separated fibres (Figure 3.2). Synthetic fibres do not show such distinctive

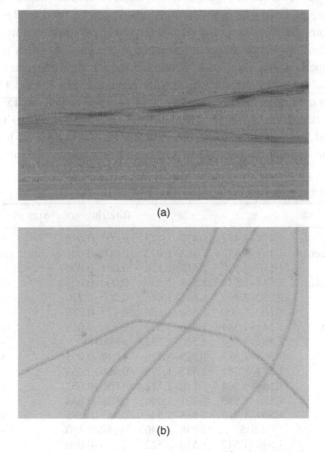

(a)

(b)

Figure 3.2 Optical microscopy of (a) cotton (×20 magnification) and (b) polyester fibres (×5 magnification).

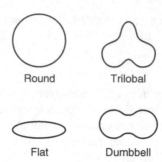

Figure 3.3 Cross-sectional shapes of fibres.

morphology (as observed in Figure 3.2), but do have various shapes based on the manufacturing process. The cross-sectional shape can be used to classify fibres, and some common cross-sectional shapes for fibres are illustrated in Figure 3.3.

A PLM can be used to gather information than can distinguish a fibre. The Becke line method may be used to determine the RI, birefringence and sign of elongation values, and typical values for fibres are listed in Table 3.1 [7]. When an anisotropic fibre is examined using a PLM at an angle of 45°, coloured interference bands are observed. The interference colour is related to the fibre diameter and birefringence using the Michel–Levy chart. If a hot stage is available, the melting temperature of synthetic fibres can be measured. Typical

Table 3.1 Optical properties of synthetic fibres

Fibre	$n_{\parallel}$	$n_{\perp}$	Birefringence	Sign of elongation
cellulose diacetate	1.474–1.479	1.473–1.477	0.002–0.005	+
cellulose triacetate	1.467–1.472	1.468–1.472	0.000–0.001	+, −
acrylic	1.510–1.520	1.512–1.525	0.001–0.005	−
modacrylic	1.538–1.539	1.538–1.539	0.000–0.001	−
Kevlar	2.050–2.350	1.641–1.646	0.200–0.710	+
Nomex	1.800–1.900	1.664–1.680	0.120–0.230	+
viscose rayon	1.541–1.549	1.520–1.521	0.020–0.028	+
lyocell	1.562–1.564	1.520–1.522	0.044	+
nylon 6	1.568–1.583	1.525–1.526	0.049–0.061	+
nylon 6,6	1.577–1.582	1.515–1.526	0.056–0.063	+
PE	1.568–1.574	1.518–1.522	0.050–0.052	+
PP	1.520–1.530	1.491–1.496	0.028–0.034	+
PET	1.699–1.710	1.535–1.546	0.147–0.175	+
PBT	1.688	1.538–1.540	0.148–0.150	+
vinal	1.540–1.547	1.510–1.522	0.025–0.030	+
vinyon	1.527–1.541	1.534–1.536	0.002–0.005	+

Table 3.2 Melting temperatures for fibres

Fibre	Melting temperature/°C
acetate	250–255
acrylic	softens 240, melts > 300
modacrylic	160
nylon 6	210–220
nylon 6,6	250–260
nylon 11	180–190
PBT	220–225
PE	120–135
PET	250–260
PP	165–180
saran	168
spandex	230
triacetate	290–300

melting temperature values for synthetic fibres that melt below 300°C are listed in Table 3.2. Solubility testing can also helpful, and common solvents include dimethylformamide and cyclohexanone.

SAQ 3.1

A synthetic fibre specimen is examined using a PLM, and the birefringence is determined to be 0.150. Is it possible to identify the fibre based on this value? What is another test that could be carried out using the microscope to assist in the identification of this fibre?

Microscopic examination of hair fibres can aid in the identification process [1–3, 9, 10]. A stereomicroscope and compound microscope can be used to characterize hair features such as the colour, the features of the shaft (e.g. scales) and the length and diameter. Figure 3.4 shows the general features of a hair. A *medullary index* (MI) can be determined using the ratio:

$$MI = \frac{\text{diameter of medulla}}{\text{diameter of hair}} \tag{3.1}$$

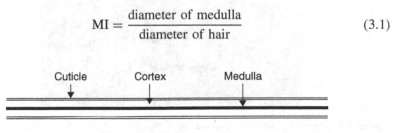

Figure 3.4 Hair structure.

Human hair shows MI values less than 0.3, while animal hair shows values ranging from 0.5 to 0.9, so human and animal hair can be differentiated using this approach.

3.2.4 Paint

The various components of paint specimens can be characterized using a number of optical microscopy techniques [2, 3, 11, 12]. The size, shape, colour, surface condition, thickness and layer sequence examined using microscopy can provide valuable information about a sample. Usually a stereomicroscope is used for initial examination. An examination of the layer structure requires that a cross-section be made using a microtome, and the specimen may need to be embedded in resin and polished to provide a sample suitable for analysis. Figure 3.5 shows an optical micrograph illustrating the layers of a cross-section of an automotive paint chip produced by embedding the chip in paraffin and sectioned using a microtome. Five separate layers can be identified for this specimen.

If enough sample is available for destructive testing, microchemical tests can be carried out. Solubility tests can be effective for the identification of different components, and there are numerous chemical tests available for the identification of the binder and pigments that can be carried out under a microscope [3, 13, 14]. Fluorescence microscopy at an excitation wavelength of 365 nm may also be used to characterize pigments and additives.

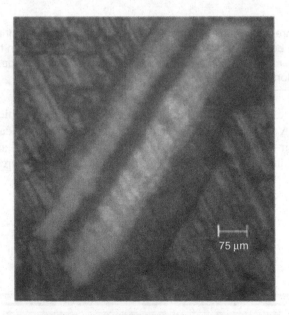

Figure 3.5 Image of paint chip cross-section. Reproduced from K. Flynn *et al.*, *J. Forensic Sci*. **50**, 832–841 (2005) with permission from John Wiley & Sons, Ltd.

3.2.5 Drugs

Drugs can be screened using microcrystalline tests, with the morphology of a crystal in precipitation tests used to identify a substance [2, 11, 15–20]. The unknown is dissolved in a solution (often aqueous), and a test reagent is added or initially contained in the solution. A reaction between the reagent and the sample forms a solid with a unique crystal structure. There is a range of established test reagents, but typically they will contain a heavy metal complexed to halogens. The reaction can simply be carried out by adding a small volume to a microscope slide. A standard light microscope is adequate for microcrystalline testing, but a PLM can be useful as some crystals show birefringence. The crystal morphology should be compared with that formed by an authenticated standard. Microcrystalline testing is commonly carried out on suspect samples. As an example, the use of two reagents (gold chloride and platinum chloride) enables amphetamine and methamphetamine to be distinguished from each other. These tests can also be utilized to determine if the d-, l- or d,l- forms of the drugs are present.

DQ 3.1

Acids, including phosphoric, sulfuric, hydrochloric and acetic acids, are commonly used as solvents in reagents for drug microcrystalline tests. Why would acids be used in such reagents?

Answer

Acids have the ability to affect the solubility of the drug crystals. For example, the addition of acetic acid produces greater solubility. Solvent properties affect the type of crystals formed.

3.2.6 Glass

Glass evidence is commonly examined using light microscopy [2, 11, 21, 22]. In the first instance, the physical appearance of small glass fragments can be examined using a stereomicroscope, and information about the type of impact can be obtained. Fluorescence microscopy can also provide discriminating information about glass evidence.

A very common method of differentiating glass fragments is the determination of the RI. If two pieces of glass have the same RI, strong evidence is provided that the glass originates from the same source. Typical RI values for headlight glass are 1.47–1.49, while values for window or bottle glass are usually in the range of 1.51–1.52. A manual Becke line method using immersion oils has been employed as a standard procedure. An automated method using commercially available Glass Refractive Index Measurement (GRIM) instrumentation has emerged and eliminates the subjectivity associated with an operator. The statistical interpretation of RI values collected for glass specimens (as well as other forms of data to be discussed in this book) can be carried out in order to determine the value of such evidence: a probabilistic or Bayesian approach involving likelihood ratios is a recognized approach [23, 24].

SAQ 3.2

Clear glass fragments are collected from the roadside adjacent to an incident involving impact between two vehicles. The RI values for the fragments are measured using a GRIM instrument, and the readings fall into a narrow range of values averaging 1.478. What is the predicted source of the glass?

3.2.7 Soil

Optical microscopy can be used to investigate the discriminating features of soil evidence [1–3, 25, 26]. An initial assessment of a soil sample can be carried out using a stereomicroscope, and contaminants such a glass or paint can be identified. The soil may be sieved into sections and the particle size distribution characterized to distinguish soils. The main classes based on diameters are:

- sand (0.05–2 mm);

- silt (0.002–0.05 mm);

- clay (< 0.002 mm).

The colour of soil can also be used to characterize a soil specimen, and a comparison with the *Munsell soil colour chart* is usually made. The Munsell chart is used to classify the colour based on hue, value and chroma. There are five primary hues (blue, green, purple, red and yellow) with five intermediate hues. The value is a measure of the darkness or lightness of the colour on a scale from 0 to 10. The chroma is the strength of the colour on a scale from 0 to 8.

PLM is used to identify soil minerals using physical appearance and optical properties including RI and birefringence. Minerals show characteristic crystal structures that can aid the identification process. Some minerals are isotropic and show one RI if they possess a cubic structure. Minerals may also be anisotropic (showing hexagonal, rhombohedral, tetragonal, rhombic, monoclinic or triclinic structures). Anisotropic minerals may be uniaxial (hexagonal, rhombohedral, tetrahedral) and have two RIs, or biaxial (rhombic, monoclinic, triclinic) and show three RIs. Table 3.3 lists the RI values for some common minerals.

SAQ 3.3

PLM is used to examine a mineral obtained from a soil specimen. A hexagonal crystal is observed, and two RIs of 1.54 and 1.55 are determined. Is it possible to identify this mineral? Is this crystal uniaxial or biaxial?

Table 3.3 Optical properties of some common minerals

Mineral	Colour	Crystal structure	RI
albite	white, grey	triclinic	1.528, 1.529, 1,536
calcite	colourless	trigonal	1.658, 1.486
dolomite	colourless	trigonal	1.682, 1.503
gypsum	colourless, white	monoclinic, rhombic	1.521, 1.523, 1.530
haematite	red, orange	trigonal	3.15, 2.87
magnetite	black, grey	cubic	2.42
malachite	green	monoclinic	1.655, 1.875, 1.909
orthoclase	colourless, white, pink	monoclinic	1.518, 1.524, 1.526
quartz	colourless	hexagonal	1.544, 1.553
rutile	white	tetragonal	2.616, 2.903

3.2.8 Documents

The stereomicroscope and comparison microscope are important tools for questioned document examination [1, 11, 27]. Handwriting properties including erasures, alterations and crossed strokes can be identified. Characteristic printing or copying defects may also be identified using this approach. Ink and toner types can be potentially discriminated using light microscopy. For example, dry and liquid toners may be distinguished. Dry toners have a raised glossy appearance, while liquid toners penetrate into the paper substrate and appear as a thin coating.

3.2.9 Firearms

A comparison microscope is a useful tool for the identification of impression evidence such as firearm and tool marks [1, 2]. Various markings can be matched and the sources of items may be determined. Adjustable holders can be used beneath the objective lenses of the microscope and both items, such as bullets, can be simultaneously observed. One item can be rotated until a distinct mark is observed, then the other item is rotated until a matching striation mark is located.

3.3 Transmission Electron Microscopy

Transmission electron microscopy (TEM) parallels optical microscopy, but high-speed electrons rather than light are used [28]. The technique exploits the diffraction of electrons, and the wave characteristics of electrons are used to obtain pictures of very small objects. Images of objects that cannot be seen with light microscopes can be produced using TEM. According to the laws of optics, it is possible to form an image of an object that is smaller than half the wavelength of the light used, which means that for visible light of wavelength at 400 nm, the smallest that may be observed is 20 μm. Charged electrons also have the advantage that they may be focused by applying an electric or magnetic field. In TEM a magnification of ×200000 and a resolution of the order of 0.5 nm may be obtained. De Broglie's equation relates wave and particle properties:

$$\lambda = \frac{h}{mv} \tag{3.2}$$

where λ is the wavelength, h is the Planck constant, m is the particle mass and v is the particle velocity. As the wavelength of an electron is inversely proportional to its velocity, by accelerating electrons at very high velocities, it is possible to obtain wavelengths as short as 0.004 nm.

3.3.1 Method

A typical layout of a transmission electron microscope is shown in Figure 3.6. An electron gun provides a source of electrons. Magnetic fields can be produced with gradients that act as lenses for the electron waves. The detector is mounted behind the sample as electrons are transmitted through a thin section of the sample. Magnification is achieved by using lenses underneath the sample to project the image formed by the transmitted electrons onto a recording device.

TEM can be combined with *energy-dispersive X-ray analysis* (EDX) (also known as energy-dispersive X-ray spectroscopy (EDS)) in order to carry out

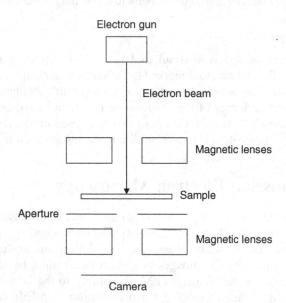

Figure 3.6 Layout of a transmission electron microscope.

elemental analysis of a specimen. Signal from the sample is converted into a voltage and sent to a converter. The data is displayed in the form of a spectrum with intensity versus emission energy. The spectrum provides a fingerprint of the specific elements present in the sample. *Electron energy loss spectroscopy* (EELS) is also a technique used in TEM and can be used to provide similar information to that given by EDS. The energy loss resulting from a beam of electrons reflected from a surface is measured and is characteristic for a material.

For sample preparation, very thin sections of the order of 50–100 nm are required for TEM – an ultramicrotome may be used for this purpose. For paint fragments, the fragment is first dehydrated by heating and then embedded in an epoxy resin. The embedded sample is then cut using a microtome to the desired thickness. Sample preparation is relatively straightforward when the sample is in the form of a powder with a small grain size, which can be dissolved in acetone or an alcohol and then a drop of the solution is placed on a metal mesh grid for analysis. For samples that are difficult to grind or where there is a risk of damage through use of a microtome, such as fibres or polymers, ion milling is a good method. In this method, the sample is bombarded with an ion or neutral atom beam that produces neutral and/or charged species in the sample.

3.3.2 Interpretation

TEM is a useful technique for characterizing structure and morphology. EDX is particularly useful for sample analysis. EDX analysis involves the analysis of the X-rays produced due to the interaction of the primary electron beam with the sample atoms. The ejection of electrons from the inner energy shells leaves a gap that is filled by an electron from an outer shell of the atom. An X-ray is released during the process that is characteristic of the atom, so the elemental composition can be determined. The intensity of the X-ray emission is plotted as a function of energy, and the positions of the observed peaks can be attributed to different elements.

Electron diffraction patterns can be used to identify crystal structures. A table of interplanar distance values is then obtained from the pattern and are compared with a crystallographic database to identify the compound.

3.3.3 Paint

TEM is a useful technique for the examination of certain paint specimens that cannot be definitively characterized using techniques such as PLM due to, say, the small size of the pigment particles [12]. The use of EDX and/or electron diffraction enables pigments and extenders to be characterized, and morphological differences can be used to discriminate specimens. TEM is regarded as useful for the examination of pearlescent pigments, which are composed of mica powders coated with TiO_2.

SAQ 3.4

As micas are sheet silicate minerals that are known to have weaker bonding between layers, what would be the recommended approach to preparing a specimen for TEM analysis?

3.4 Scanning Electron Microscopy

Scanning electron microscopy (SEM) is also a technique that utilizes an electron beam to produce magnified images of samples [29, 30]. SEM is used more widely for forensic purposes than TEM. The sample surface is scanned with a beam of energetic electrons, and a number of processes occur that can be used to generate an image of the sample surface. These include secondary electrons (SEs), backscattered electrons (BSEs) and the production of characteristic X-rays. SEM is commonly combined with EDX in order to carry out elemental analysis.

3.4.1 Methods

A scanning electron microscope operates by producing a beam of electrons in a vacuum. Figure 3.7 shows a schematic diagram of a scanning electron microscope. An electron beam is generated by an electron gun, and electromagnetic lenses are used to focus the beam. The beam is swept back and forth, known as rastering, over a selected area of the sample. The interaction of the electrons with the sample causes electrons to be dislodged from the atoms within the sample. The electrons generated are detected, amplified and displayed.

Sample preparation for SEM is critical for obtaining quality results from the technique. SEM specimens are mounted on a metal disc known as a stub. Conducting adhesives or adhesive tapes can be used to adhere the sample to the stub. Embedding samples in resin is a useful approach, particularly for layered specimens such as paints, when a flat surface is required for EDX analysis. The embedded material is cut to expose a cross-section and polished to a smooth surface using an abrasive such as diamond paste. If the embedding approach is not appropriate due to the need to retrieve the sample for further analysis, the sample can be mounted on the stub to expose the cross-section to the electron beam. There is also a stair-step approach available that involves exposing layers using a scalpel.

For nonconductive samples it is necessary to coat the surface of the sample with a thin conducting layer. If a nonconductive sample is examined, secondary electrons leaving the sample generate an excess of positive charge of the surface and result in a blurred image. The layer is applied using a vacuum evaporator (known as sputter coating), and gold and carbon are commonly used deposition materials. If EDX analysis is to be carried out, then care must be exercised

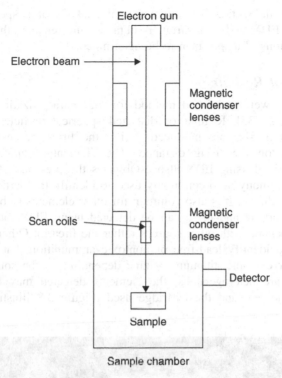

Figure 3.7 Schematic diagram of a scanning electron microscope.

in the choice of coating material so that the coating material does not interfere with the elemental analysis. *Environmental scanning electron microscopy* (ESEM) can be used in order to avoid the coating of samples and for sensitive samples that may not be stable in a vacuum. ESEM uses a moderate pressure of a gas such as water in the sample chamber instead of a sample coating to remove the excess charge.

3.4.2 Interpretation

SEs are ejected from atoms during the interaction with the electron beam and escape from the sample near the surface. These electrons provide information about the topography of the sample surface. BSEs are produced when electrons are elastically scattered by the sample atoms on interaction with the electron beam. Such electrons are more energetic than SEs and are produced from deeper in the samples. The images produced by BSEs show poorer resolution compared to SEs, but the images can be correlated with the composition. A BSE image is brighter in regions containing higher atomic numbers.

As described in Section 3.3, an elemental analysis of a specimen can be obtained using EDX and a spectrum is obtained that enables the composition of different regions of a specimen to be determined.

3.4.3 Gunshot Residue

SEM–EDX is a well-established method for the characterization of inorganic GSR particles [29–35]. Regular or distorted spherical particles of the order of micrometres in size are produced during the firing process, the size of which depends on the firing distance. The elemental composition of the particles determined using EDX also establishes the presence of GSR, with a lead–barium–antimony ratio commonly used to identify the particles that result from firearm discharge. It is also common for other elements to be incorporated into the GSR particle and these can be detected using EDX, but they can be potentially associated with other sources rather the firearm. Other elements can be monitored to identify lead-free or nontoxic ammunition that can show the presence of, for example, titanium or zinc depending on the source. Although GSR particles are heterogeneous, the elements detected may be able to be connected to the gun and the cartridge used. Figure 3.8 illustrates a typical

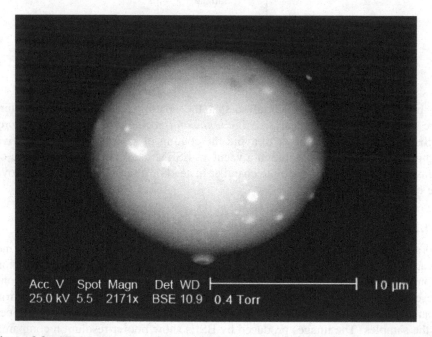

Figure 3.8 SEM image of a GSR particle (after shooting a Winchester 9 mm Luger cartridge). Reproduced from M. Morelato *et al.*, *Forensic Sci. Int.* **217**, 101–106, 0379-0738 with permission from Elsevier (2012).

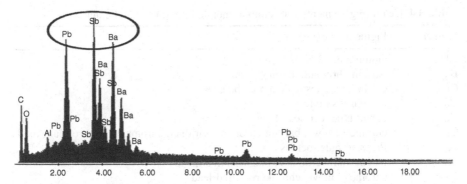

Figure 3.9 EDX spectrum of the GSR particle shown in Figure 3.8. Reproduced from M. Morelato *et al.*, *Forensic Sci. Int.* **217**, 101–106, 0379-0738 with permission from Elsevier (2012).

electron micrograph of a GSR particle detected from the BSE intensity, and Figure 3.9 shows the corresponding EDX spectrum of this particle.

SAQ 3.5

Would it better to use an aluminium tape or a carbon tape for the collection of GSR for examination by SEM–EDX?

3.4.4 Paint

The morphology and composition of paint samples are commonly examined using SEM–EDX [12, 30, 36]. Paint is generally embedded in resin for SEM analysis and sections of the order of tens of micrometres are recommended to obtain data of suitable quality. The features of individual layers of multilayer paint specimens can be observed using SEs and BSEs. EDX allows the elemental composition of paint layers to be determined. Particular elements can indicate the presence of specific inorganic pigments and extenders and these are summarized in Table 3.4 [12, 30]. Due to the heterogeneous nature of paint specimens, data are often compared with reference data or the ratios of peaks are used. Mapping of the elemental composition also enables more reliable information to be obtained.

SAQ 3.6

SEM–EDX analysis is carried out on a yellow coloured paint pigment particle. The EDX spectrum indicates the presence of chromium, lead and sulfur. What type of pigment is likely to be present?

Table 3.4 Identifying elements for common inorganic pigments and extenders

Element	Pigment or extender
Al	aluminium silicates
Ba	barium chromate, barium sulfates
Ca	calcium sulfates, calcium carbonates
Cd	cadmium sulfate
Co	cobalt blue, cerulean blue
Cr	chrome yellows, barium chromate, viridian, chrome reds, chromium oxide
Fe	Prussian blue, ochres
Mg	magnesium carbonate, magnesium silicate
Pb	chrome yellows, chrome reds, red lead
S	cadmium sulfides, chrome yellows, calcium sulfate, barium sulfates
Si	ochres, aluminium silicates, magnesium silicate

3.4.5 Fibres

The high magnification and greater depth of field provided by SEM make this technique a useful tool for studying the surface morphology of fibres [6, 37, 38]. The features of a fibre surface, ends or cross-section can provide valuable identifying information. The value of SEM for forensic fibre analysis lies in its combination with EDX. Discrimination of fibres is feasible via the characterization of the inorganic additives using elemental analysis.

SAQ 3.7

A blue textile fibre is examined using SEM–EDX, and Fe is found to be present. If the Fe is associated with the pigment used to colour the fibre, is it possible to identify the pigment used?

3.4.6 Documents

SEM can be applied to the characterization of toners used in questioned documents [27]. The surface morphology of toners is observed, and it can be determined what type of toner fusion has taken place, thus providing information that can be used to differentiate toners. SEM–EDX provides information about the toner composition. The magnetic components of toners, such as ferrite or magnetite, can be identified. The following elements may be detected in toners and enable minor components to be used to provide further discrimination: Ca, Cl, Cr, Cu, Fe, Mn, Ni, S, Si, Sr and Zn. It is important to also analyse the paper substrate as paper fillers can contain elements observed in toners.

SEM is also useful for the analysis of line crossings in documents and can be employed where optical microscopy does not provide the necessary resolving power [40]. SEM works well where the ink is not absorbed into the paper fibres.

SEM allows for the morphological properties to be visualized due to the resolving power and depth of focus.

DQ 3.2

Why should optical microscopy be used prior to any SEM examination of a document?

Answer

As paper is a nonconductive material, coating the specimen is most likely required for successful SEM analysis. Optical microscopy should be carried out before such modification is carried out.

3.4.7 Glass

SEM–EDX can be used to characterize small fragments of glass in a nondestructive manner with minimal sample preparation involved [32, 39, 41–43]. The technique is limited to major and minor elements such as O, Na, Ca, Fe, Al, Mg, Si and K rather than trace elements due to the improved sensitivity for lower atomic number elements. Elemental ratios, such as the Ca/Mg ratio, can be employed in the classification process when compared to glass standards.

3.5 Atomic Force Microscopy

Atomic force microscopy (AFM) is used to examine the surface topography of samples at the atomic level [44]. The technique is based on the idea that an electron in an atom has a small probability of existing far from the nucleus and, under certain circumstances, it can tunnel and appear closer to another atom. The tunnelling electrons create a current that can be used to image the atoms of an adjacent surface.

3.5.1 Methods

AFM can be used to examine both conducting and insulating surfaces. A flexible cantilever with a tip attached is scanned in a raster pattern over a sample surface (Figure 3.10). The force between the surface and the cantilever causes the cantilever to be deflected and the small movements are optically measured. During a scan, the force on the tip can be held constant by the tip motion and topographic information is provided by the experiment. The common operating modes of AFM are contact mode and tapping mode. In contact mode the tip is in direct contact with the sample surface and is applicable for samples that are tightly attached to a substrate. In tapping mode the tip is not in contact with the surface, but taps it gently and is useful for samples not strongly fixed to a substrate (e.g. adhesives).

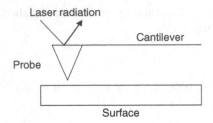

Figure 3.10 Atomic force microscopy (AFM).

3.5.2 *Interpretation*

The interaction force between the tip (F) and the sample in AFM can be directly obtained from the cantilever deflection using Hooke's law:

$$F = -kx \tag{3.3}$$

where k is the force constant and x is the deflection. Commonly the surface topography of surfaces is imaged using AFM, but AFM is also a tool for sensitive force measurements. A study of the interaction forces can lead to a clearer understanding of processes at a molecular level.

3.5.3 *Documents*

The high resolving power and ability to examine documents without surface modification mean that AFM has great potential in document examination [45]. Figure 3.11 shows an AFM image that illustrates how ink deposits on paper may be investigated using this technique. This shows a line crossing on paper produced with a dot matrix printer and a ballpoint pen and demonstrates that the ribbon deposit is covered by ballpoint ink in this case.

AFM is a relatively new technique, and thus far, its application to forensic problems has been limited. However, with the expansion of AFM instrumentation and the development of portable machines enabling samples of various sizes to be examined, this method provides great potential as a sensitive technique for discriminating difficult samples. Preliminary studies have demonstrated the applicability of AFM to the examination of hair, fingerprints, fibres and bloodstains.

3.6 X-Ray Diffraction

X-ray diffraction (XRD) is a technique used to determine the arrangement of atoms in solids and is useful for identifying the crystalline components of a range of materials [46, 47]. The wavelengths of certain X-rays are approximately equal

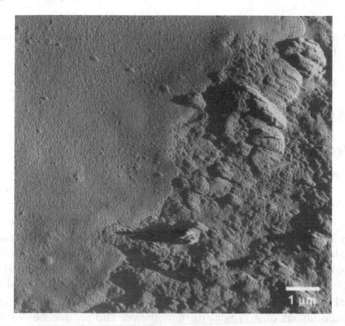

Figure 3.11 AFM image of a line crossing of ballpoint pen (left) and dot matrix deposit (right). Reproduced from S. Kasas, A. Khanmy-Vital and G. Dietler, 'Examination of line crossings by atomic force microscopy', *Forensic Sci. Int.* **119**, 3, 290–298 with permission from Elsevier (2001).

to the distance between the planes of atoms in crystalline solids, so reinforced diffraction peaks of radiation with varying intensity are produced when a beam of X-rays strikes a crystalline solid (illustrated in Figure 3.12). Two waves reach the crystal at an angle θ and are diffracted at the same angle by adjacent layers. When the first wave strikes the top layer and the second wave strikes the next layer, the waves are in phase. Constructive interference of the rays occurs at an

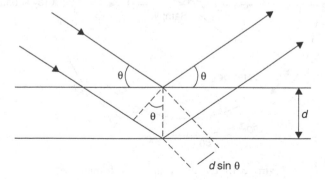

Figure 3.12 X-ray diffraction.

angle θ if the pathlength difference is equal to a whole number of wavelengths, $n\lambda$, where n is an integer and λ is the wavelength. The fundamental relationship in XRD is the Bragg equation:

$$n\lambda = 2d \sin \theta \qquad (3.4)$$

where d is the distance between the layers of atoms.

3.6.1 Methods

X-ray radioactive sources and rotating anode tubes can be used as X-ray sources in XRD. Cr, Fe, Cu or Mo may be used as anode materials. A commonly used XRD technique is the powder method, where a monochromatic X-ray beam is directed at a powdered sample spread on a support. It is important that homogenous samples are prepared for this technique in order to provide satisfactory results. At least a few mg of sample is required, but the more recent development of micropowder diffraction sampling aids in forensic applications. The single-crystal diffraction technique may also be applied to very small samples, with a tiny crystal of about 0.1 mm per side required for this approach. Figure 3.13 illustrates the layout of a typical diffractometer. The sample holder is able to be rotated. The intensities of the diffracted beams are recorded by a detector mounted on a movable carriage, which can also be rotated, and the angular position is measured in terms of 2θ, the diffraction angle. As the detector moves at a

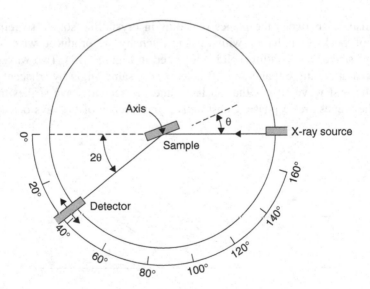

Figure 3.13 Layout of an X-ray diffractometer.

constant velocity, a computer plots the diffracted beam intensity that is collected as a function of 2θ.

3.6.2 Interpretation

The XRD pattern obtained is characteristic of the material under study, and comparison with a database of XRD patterns, such as that developed by the International Centre for Diffraction Data (ICDD), allows the material to be identified. The peaks appearing in XRD patterns can be correlated to crystalline phases with a specific structure. The evaluation of mixtures can be complicated by the shifting of values, so care must be taken when matching peaks. Qualitative analysis is usually carried out for forensic applications, but quantitative analysis is feasible. A drawback of powder diffraction is that the patterns of the constituent phases of a mixture strongly overlap, making quantitative analyses more complex. However, the Rietveld refinement is a method that has been developed which separates the individual constituents of a XRD pattern.

3.6.3 Explosives

XRD allows the composition of pre- and post-blast explosive mixtures to be determined [47, 48]. The identification of minor discriminating components can be made using this approach. Reaction products, as well as the original components, can be identified using XRD analysis. An additional benefit of application of this technique to explosives analysis is the ability to discriminate isomers of organic explosives.

3.6.4 Paint

XRD is widely used to identify inorganic pigments in paints and can be employed to confirm the identity of paint evidence or differentiate paint specimens that appear visually similar [47–49]. This technique is useful for differentiating pigments of the same chemical structure, but with different crystalline phases: such differences can be readily detected in XRD data. For example, the white pigments rutile and anatase are two different morphological forms of titanium dioxide. The XRD patterns illustrate notable differences (Figure 3.14) [50]. The significance of matching certain paints must be placed in context as some pigments are very common. For example, TiO_2 in the form of rutile is a common single pigment component in white paints and should not necessarily be used in isolation to match paint evidence.

DQ 3.3

Why is XRD a more suitable technique for inorganic pigments rather than organic pigments?

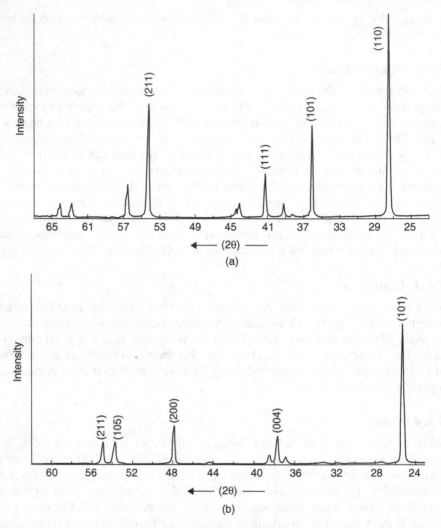

Figure 3.14 XRD patterns of rutile and anatase. Reproduced from M. Kotrlý, 'Application of x-ray diffraction in forensic science', *Z. Kristallograph. Suppl.* **23**, 35-40, with permission from Elsevier (2006).

Answer

Inorganic pigments are more likely to produce the crystalline structures that are characterized by XRD. Organic pigments do not tend to show such morphology, and so are more difficult to characterize using XRD.

3.6.5 Drugs

Although not necessarily the first choice of technique for drug analysis, XRD can be employed to identify drug components with reference to an appropriate database [47, 49, 51]. XRD is usually used to characterize the structural form of a drug molecule, for instance whether it is a salt, base or acid. Diluents such as sugars or adulteration may be confirmed using this approach. A comparison of seizures is feasible with XRD data: the patterns of additives of reaction by-products can be used to find links.

SAQ 3.8

XRD analysis is to be carried out on a white powder suspected of being cocaine. What sort of compounds might be expected to be detected by XRD?

3.6.6 Documents

XRD can provide information regarding paper composition for document examination [47, 49, 52]. The main application is the identification of the filler. Filler materials, such as TiO_2 and kaolinite, have crystalline structures that produce clear diffraction patterns. Filler types can be discriminated, enabling paper sample to be linked to a source. It is also possible to investigate the cellulose structure of paper using XRD. The degree of crystallinity of cellulose varies depending on the raw material and processing conditions, and so can be used to discriminate documents.

3.6.7 Soil

XRD is an established technique for soil analysis [47]. The commonly observed minerals found in soil, such as quartz, clays, feldspar and calcite, produce distinct

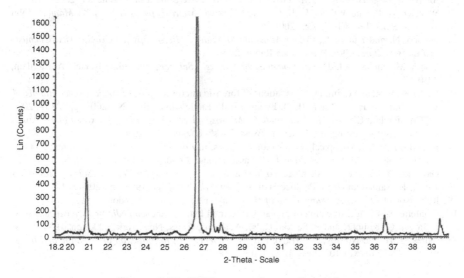

Figure 3.15 XRD pattern for a soil sample.

XRD patterns that enable soils to be characterized and compared. The XRD pattern obtained for a grave soil is shown in Figure 3.15. The mineral content can be identified by comparison with a XRD mineral database and, in this case, quartz, calcite, illite, orthoclase and albite are the minerals identified.

3.7 Summary

The use of optical microscopy is often one of the first techniques used in forensic science for the identification and characterization of evidence. A broad range of specimen types can be studied, including fibres, paint, drugs, glass, soil, questioned documents and firearms. The physical and chemical properties of specimens can be examined with optical microscopy for such purposes. TEM provides a higher resolution technique that enables samples, such as paints, to be investigated. SEM enables a broader range of specimens, such as gunshot residue, paint, fibres, questioned documents and glass, to be examined to a greater resolution and also enables elemental analysis of evidence to be carried out. AFM is a microscope technique that shows great promise for forensic applications such as in document examination. The microstructural properties of crystalline materials, including explosives, paint, drugs, questioned documents and soil, can also be examined using XRD.

References

1. Wilson, L. J. and Wheeler, B. P., Optical microscopy in forensic science, in *Encyclopedia of Analytical Chemistry*, R. A. Meyers (Ed.), John Wiley & Sons, Ltd, Chichester, 2009.
2. Wheeler, B. P., and Wilson, L. J., *Practical Forensic Microscopy: A Laboratory Manual*, John Wiley & Sons, Ltd, Chichester, 2008.
3. Petraco, N. and Kubic, T., *Colour Atlas and Manual of Microscopy for Criminalists, Chemists and Conservators*, CRC Press, Boca Raton, 2004.
4. Houck M. and Siegel, J., *Fundamentals of Forensic Science*, 2nd ed., Elsevier, Amsterdam, 2010.
5. DeForest, P. R. and Grim, D., Foundations of forensic microscopy, in *Forensic Science Handbook Vol.* 1, 2nd ed., R. Saferstein (Ed.), Prentice Hall, Englewood Cliffs, NJ, 2002, pp. 215–320.
6. ASTM Standard E2228, *Standard Guide for Microscopical Examination of Textile Fibres*, American Society for Testing and Materials, West Conshohocken, PA, 2010.
7. Palenik, S. J., Microscopical examination of fibres, in *Forensic Examination of Fibres*, 2nd ed., J. Robertson and M. Grieve (Eds), Taylor and Francis, London, 1999, pp. 153–178.
8. Gaudette, B. D., The forensic aspects of textile fibre examination, in *Forensic Science Handbook Vol.* 2, R. Saferstein (Ed.), Prentice Hall, Englewood Cliffs, NJ, 1988, pp. 209–272.
9. Robertson, J., *Forensic Examination of Hair*, Taylor and Francis, London, 1999.
10. Gaudette, B. D., Identification of human and animal hair, in *Encyclopedia of Forensic Sciences*, J. Siegel, G. Knupfer and P. Saukko (Eds), Academic Press, New York, 2000, pp. 1034–1041.
11. James, S. H. and Nordby, J. J. (Eds), *Forensic Science: An Introduction to Scientific and Investigative Techniques*, CRC Press, Boca Raton, 2003.

12. Buzzini, P. and Stoecklein, W., Forensic sciences / paints, varnishes and lacquers, in *Encyclopedia of Analytical Science*, P. Worsfield, A. Townshend and C. F. Poole (Eds), Elsevier, Amsterdam, 2005, pp. 453–464.

13. Fiegl, F., *Spot Tests in Inorganic Analysis*, 6th ed., Elsevier, Amsterdam, 1988.

14. Fiegl, F. and Anger, V., *Spot Tests in Organic Analysis*, 7th ed., Elsevier, Amsterdam, 1966.

15. Elie, M. P. and Elie, L. E., Microcrystalline tests in forensic drug analysis, in *Encyclopedia of Analytical Chemistry*, R. A. Meyer (Ed.), John Wiley & Sons, Inc., New York, 2009.

16. Fulton, C., *Modern Microcrystal Tests for Drugs*, John Wiley & Sons, Inc., New York, 1969.

17. Levine, B., Presumptive chemical tests, in *Encyclopedia of Forensic Sciences*, J. Siegel, G. Knupfer and P. Saukko (Eds), Academic Press, New York, 2000, pp. 167–172.

18. Siegel, J. A., Forensic identification of controlled substances, in *Forensic Science Handbook Vol. 2*, R. Saferstein (Ed.), Prentice Hall, Englewood Cliffs, 1988, pp. 68–160.

19. ASTM Standard E1969, *Standard Guide for Microcrystal Testing in the Forensic Analysis of Methamphetamine and Amphetamine*, American Society for Testing and Materials, West Conshohocken, PA, 2006.

20. ASTM Standard E1968, *Standard Guide for Microcrystal Testing in the Forensic Analysis of Cocaine*, American Society for Testing and Materials, West Conshohocken, PA, 1998.

21. Hamer, P. S., Microscopic techniques for glass examinations, in *Forensic Examination of Glass and Paint Analysis and Interpretation*, B. Caddy (Ed.), Ellis Horwood, London, 2001.

22. Koons, R., Buscaglia, J., Bottrell M. and Miller, E. T., Forensic glass comparisons, in *Forensic Science Handbook Vol. 1*, 2nd ed., Prentice Hall, Englewood Cliffs, NJ, 2002, pp. 161–214.

23. Daeid, N. N., Statistical interpretation of glass evidence, in *Forensic Examination of Glass and Paint*, B. Caddy (Ed.), Taylor and Francis, London, 2001.

24. Zadora, G., Chemometrics and statistical considerations in forensic science, in *Encyclopedia of Analytical Chemistry*, R. A. Meyers (Ed.), John Wiley & Sons, Ltd, Chichester, 2010.

25. Petraco, N., *American Lab* **26**, 35–40 (1994).

26. Murray, R. C. and Solebello, L. P., Forensic examination of soil, in *Forensic Science Handbook Vol. 1*, 2nd ed., Prentice Hall, Englewood Cliffs, NJ, 2002, pp. 615–634.

27. Neumann, C. N. and Mazzella, W. D., Forensic sciences / questioned documents, in *Encyclopedia of Analytical Science*, 2nd ed., P. Worsfold, A. Townshend and C. Poole (Eds), Elsevier, Amsterdam, 2005, pp. 465–471.

28. Flegler, S. L., Heckman, J. W. and Klompaiens, K. L., *Scanning and Transmission Electron Microscopy: An Introduction*, Oxford University Press, Oxford, 1995.

29. Basu, S., Scanning electron microscopy in forensic science, in *Encyclopedia of Analytical Chemistry*, R. A. Meyers (Ed.), John Wiley & Sons, Inc., New York, 2006.

30. Henson, M. L. and Jergovich, T. A., Scanning electron microscopy and energy dispersive X-ray spectrometry (SEM/EDS) for the forensic examination of paints and coatings, in *Forensic Examination of Glass and Paint*, B. Caddy (Ed.), Taylor and Francis, London, 2001, pp. 243–272.

31. ASTM Standard E1588, *Standard Guide for Gunshot Residue Analysis by Scanning Electron Microscopy/Energy Dispersive X-ray Spectrometry*, American Society for Testing and Materials, West Conshohocken, PA, 2010.

32. Zadora G. and Brozek-Mucha, Z., *Material Chemistry and Physics* **81**, 345–348 (2000).

33. Schwoeble, A. J. and Exline, D. L., *Current Methods in Forensic Gunshot Residue Analysis*, CRC Press, Boca Raton, 2000.

34. Dalby, O. Butler, D. and Burkett, J. W., *J. Forensic Sci.* **55**, 924–943 (2010).

35. Romolo, F. S. and Margot, P., *Forensic Sci. Int.* **11**, 195–211 (2001).

36. ASTM Standard E1610, *Standard Guide for Forensic Paint Analysis and Comparison*, American Society for Testing and Materials, West Conshohocken, PA, 2002.

37. Roux, C., Instrumental methods used in fibre examination 9.2: Scanning electron microscopy and elemental analysis, in *Forensic Examination of Fibres*, 2nd ed., J. Robertson and M. Grieve (Eds), Taylor and Francis, London, 1999.

38. Greaves, P. H. and Saville, B. P., Scanning electron microscopy, in *Microscopy of Textile Fibres*, Bios Scientific Publishers, Oxford, 1995, pp. 51–67.
39. Lambert, J. A., Forensic sciences / glass, in *Encyclopedia of Analytical Science*, 2nd ed., P. Worsfold, A. Townshend and C. Poole (Eds), Elsevier, Amsterdam, 2005, pp. 423–430.
40. Waeschle, P. A., *J. Forensic Sci.* **24**, 569–578 (1979).
41. Almirall, J. A., Elemental analysis of glass fragments, in *Forensic Examination of Glass and Paint*, B. Caddy (Ed.), Taylor and Francis, London, 2001.
42. Falcone, R., Sommariva, G. and Verita, M., *Microchim. Acta* **155**, 137–140 (2006).
43. Buscaglia, J. A., *Anal. Chim. Acta* **288**, 17–24 (1994).
44. Eaton, P. and West, P., *Atomic Force Microscopy*, Oxford University Press, Oxford, 2010.
45. Kasas, S., Khanmy-Vital, A. and Dietler, G., *Forensic Sci. Int.* **119**, 290–298 (2001).
46. Janssens, K., X-ray based methods of analysis, in *Non-Destructive Microanalysis of Cultural Heritage Materials*, K. Jannsens and R. Van Grieken (Eds), Elsevier, Amsterdam, 2004, pp. 129–226.
47. Fischer, R. and Hellmiss, G., Principles and forensic applications of X-ray diffraction and X-ray fluorescence, in *Advances in Forensic Sciences*, *Vol.* 2, H. C. Lee and R. E. Gaensslen (Eds), Year Book Medical Publishers, Chicago, 2003, pp. 129–158.
48. Kotrlý, M., *Z. Kristallograph. Suppl.* **23**, 35–40 (2006).
49. Rendle, D. F., *Rigaku Journal*, **19**, 11–22 (2003).
50. Debnath, N. C., and Vaidya, S.A. *Prog. Org. Coatings* **56**, 159–168 (2006).
51. Rendle, D. F., Forensic science: Every contact leaves a trace, in *Industrial Applications of X-ray Diffraction*, F. H. Chung and D. K. Smith (Eds), Marcel Dekker, New York, 2000, pp. 659–676.
52. Ellen, D., *The Scientific Examination of Questioned Documents: Methods and Techniques*, Taylor and Francis, London, 1997.

Chapter 4

Molecular Spectroscopy

Learning Objectives

- To understand the origins of infrared spectroscopy and how to obtain infrared spectra for forensic samples.
- To apply infrared spectroscopy to the study of paint, fibres, polymers, documents, explosives and drugs.
- To understand the origins of Raman spectroscopy and how to obtain Raman spectra for forensic samples.
- To apply Raman spectroscopy to the study of drugs, paint, fibres, documents and explosives.
- To understand the origins of ultraviolet–visible (UV-visible, or UV–vis) spectroscopy and how to obtain UV–vis spectra for forensic samples.
- To apply UV–vis spectroscopy to the study of fibres, paint, documents, drugs and toxicological samples.
- To understand the origins of fluorescence spectroscopy and how to obtain fluorescence spectra for forensic samples.
- To apply fluorescence spectroscopy to the study of body fluids and toxicological samples.
- To understand the origins of nuclear magnetic resonance spectroscopy and how to obtain nuclear magnetic resonance spectra for forensic samples.
- To apply nuclear magnetic resonance spectroscopy to the study of drugs and explosives.

Forensic Analytical Techniques, First Edition. Barbara Stuart.
© 2013 John Wiley & Sons, Ltd. Published 2013 by John Wiley & Sons, Ltd.

4.1 Introduction

Spectroscopy involves the analysis of electromagnetic radiation absorbed, emitted or scattered by molecules or atoms as they undergo transitions between energy levels. According to quantum theory, molecules and atoms exist in discrete states known as energy levels. The frequency (ν) of the electromagnetic radiation associated with a transition between two energy levels (ΔE) is given by $\Delta E = h\,\nu$, where h is the Planck constant. Different types of spectroscopic techniques using radiation from different regions of the electromagnetic spectrum are used to investigate different magnitudes of energy level separation. Atomic spectroscopy provides information about the electronic structure of an atom and is described in Chapter 5. Molecular spectroscopy is more complex and involves rotational, vibrational and electronic transitions. The spectral frequencies associated with each molecule mean that molecular spectroscopy can be used to characterize specimens. The forms of molecular spectroscopy that can be utilized for forensic analysis are infrared, Raman, ultraviolet–visible (UV–vis), fluorescence and nuclear magnetic resonance spectroscopies.

4.2 Infrared Spectroscopy

Infrared spectroscopy is a technique based on the vibrations within a molecule [1–4]. An infrared spectrum is obtained by passing infrared radiation through a sample or reflecting radiation from the sample surface and then determining what fraction of the incident radiation is absorbed at a particular energy. The energy at which a band in an absorption spectrum appears corresponds to the frequency of the vibration of a part of a molecule. For a molecule to show infrared absorptions it must possess a specific feature: an electric dipole moment of the molecule must change during the vibration. This is known as the selection rule for infrared spectroscopy.

4.2.1 Methods

Fourier transform infrared (FTIR) spectrometers are the most regularly used instruments for recording infrared spectra. The layout of an FTIR spectrometer is shown in Figure 4.1. As the most useful spectral region is the mid-infrared spectrum ($4000 - 400\,\text{cm}^{-1}$), most instruments are set up to record in this range. Radiation emerging from a source is passed through an interferometer, which produces a signal that can be mathematically transformed to rapidly produce the required spectral data. From the interferometer, the radiation interacts with the sample before reaching a detector (deuterium triglycine sulfate (DTGS) or liquid N_2–cooled mercury cadmium telluride (MCT) are popular). In addition to laboratory-based instruments, portable handheld infrared spectrometers are commercially available and provide valuable tools for fieldwork.

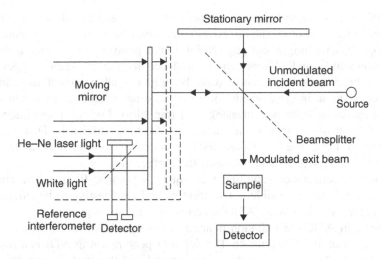

Figure 4.1 Layout of an infrared spectrometer.

It is possible to combine an infrared spectrometer with a microscope facility in order to study very small samples. In *FTIR microscopy*, the microscope sits above the FTIR sampling compartment. Infrared radiation from the spectrometer is focused onto a sample placed on a standard microscope *x-y* stage. After passing through the sample, the infrared beam is collected by a Cassegrain objective that produces an image of the sample within the barrel of the microscope, and a variable aperture is placed in this image plane. The radiation is then focussed on a small area MCT detector by another Cassegrain condenser. The microscope also contains glass objectives to allow visual inspection of the sample. By switching mirrors in the optical set-up, the microscope can be converted from transmission mode to reflectance mode. *Infrared imaging* using FTIR microspectroscopic techniques is emerging as an effective approach to studying complex specimens. The technique can be used to produce a two- or three-dimensional picture of the properties of a sample. A large number of detector elements are read during the acquisition of spectra, and this allows thousands of interferograms to be collected simultaneously and then transformed into infrared spectra.

Importantly, infrared spectroscopy enables a range of sample types to be investigated. *Transmission* spectroscopy, where the absorption of radiation is measured after it passes through a sample, is the traditional sampling method. Liquids, solids, or gases can be studied using this approach. Liquid cells commonly use alkali halide (e.g. potassium bromide, KBr) windows and a PTFE spacer, available in a variety of thicknesses. These cells are filled using a syringe, and the syringe ports are sealed with PTFE plugs before sampling.

Solids can be examined contained in KBr discs. Powdered samples (about 2–3 mg) are mixed with KBr (about 200 mg) and ground in an agate mortar and pestle before being subjected to pressure in an evacuated die to produce a transparent disc. Solids (about 50 mg) may also be mixed and ground with 1–2 drops of a mulling agent (e.g. Nujol–liquid paraffin) to produce a smooth paste known as a mull. Films can be produced for certain sample types (e.g. polymers) by either solvent casting (dissolving in a suitable solvent and allowing to evaporate on an infrared window) or melt casting (heating a specimen above the melting temperature and pressing to a thin film). For solid specimens that are too thick to measure via transmission, a diamond anvil cell (DAC) can be employed. A DAC uses two diamonds to compress a sample to a thickness suitable for measurement and increases the surface area.

The use of reflectance sampling techniques is popular for forensic applications because these can enable nondestructive and field testing to be carried out. *Attenuated total reflectance* (ATR) spectroscopy involves pressing a crystal on the sample. In ATR, the radiation penetrates the sample surface and selectively absorbs radiation in close contact. The depth of penetration in ATR is a function of wavelength, the refractive index of the crystal and the angle of incident radiation, so the relative intensities of the infrared bands in an ATR spectrum will appear different to those observed in a transmission spectrum: the intensities will be greater at lower wavenumber values. The crystals used in ATR cells are made from materials that possess low water solubility and are of a very high refractive index. Such materials include ZnSe, Ge and a thallium bromide–thallium iodide blend (KRS-5). Figure 4.2 illustrates a micro-ATR accessory.

Diffuse reflectance spectroscopy, also regularly called DRIFT (diffuse reflectance infrared by Fourier transform), is a useful technique for the examination of powdered specimens. Diffuse reflectance results from energy that penetrates one or more particles, and is reflected in all directions. Some samples can be examined as pure samples, but a powdered sample is usually mixed with KBr powder (1–5 w %). The technique avoids the need to press pellets. Diffusely scattered light can also be collected directly from a sample surface by using an abrasive diamond or graphite sampling pad. The resulting spectra can appear different from the transmission equivalent (e.g. stronger than

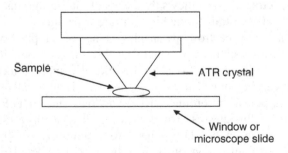

Figure 4.2 Micro ATR accessory.

expected absorption from weak bands), but the Kubelka–Munk equation can be applied to compensate for these differences.

SAQ 4.1

Which infrared sampling method would be the best choice for the following sample types?

(a) a powdered drug.

(b) a single textile fibre.

(c) a painted wall surface.

4.2.2 Interpretation

A spectrum is produced in infrared spectroscopy and can show % transmittance, absorbance or % reflectance as a function of wavenumber with units of cm^{-1}. Absorbance values should be used for quantitative analysis.

The bands appearing in an infrared spectrum can each be assigned to the vibrational modes for the sample under study. For a molecule to show infrared absorptions, it must possess a specific feature: an electric dipole moment of the molecule must change during the vibration. This is the selection rule for infrared spectroscopy. Vibrations can involve either a change in bond length (stretching) or bond angle (bending). Some bonds can stretch in-phase (symmetrical stretching) or out-of-phase (asymmetric stretching). If a molecule has different terminal atoms, then the two stretching modes are no longer symmetric and asymmetric vibrations of similar bonds, but will have varying proportions of the stretching motion of each group known as coupling. The complexity of an infrared spectrum arises from the coupling of vibrations over a large part of (or over the complete) molecule, and such vibrations are called skeletal vibrations. Bands associated with skeletal vibrations are likely to conform to a pattern or fingerprint of the molecule as a whole, rather than a specific group within the molecule.

An infrared spectrum can be divided into three main regions – the far-infrared (< 400 cm^{-1}), the mid-infrared ($4000 - 400$ cm^{-1}) and the near-infrared ($13000 - 4000$ cm^{-1}) – but the mid-infrared spectrum is the most valuable for forensic applications. The mid-infrared spectrum can be sub-divided into the X–H stretching region ($4000 - 2500$ cm^{-1}), the triple bond region ($2500 - 2000$ cm^{-1}), the double bond region ($2000 - 1500$ cm^{-1}) and the fingerprint region ($1500 - 600$ cm^{-1}), based on the nature of the functional groups present in the molecular structure. Table 4.1 summarizes the principal mid-infrared bands that may be encountered when examining materials in forensic science.

Table 4.1 Common infrared spectral bands

Wavenumber (cm^{-1})	Assignment
3700 – 3600	O—H stretching
3400 – 3300	N—H stretching
3100 – 3000	aromatic C—H stretching
3000 – 2850	aliphatic C—H stretching
2300 – 2050	C≡C stretching
2300 – 2200	C≡N stretching
1830 – 1650	C=O stretching
1650	C=C stretching
1500 – 650	fingerprint region (bending, rocking)

Detailed correlation tables are widely available that allow bands to be assigned to a particular vibrational mode and, therefore, the spectrum attributed to a particular molecular structure. Spectral library databases are available for materials of forensic interest and allow an unknown to be matched based on the intensities and wavenumber values of bands [5]. Commercial forensic infrared databases are available for the identification of materials including drugs, polymers and paint and fibre components. Where a straightforward identification is not able to be made by comparison with a library database, multivariate statistical methods (outlined in Chapter 1), such as PCA, can be applied [6].

4.2.3 Paint

Infrared spectroscopy is a valuable technique for the analysis of paint specimens as it is capable of providing structural information about both the inorganic and organic components found in paint [1–3, 7–9]. A number of sampling approaches to forensic paint analysis use FTIR spectroscopy. If no microscope is available, layer separation by scalpel or a microtome followed by analysis using a KBr disc in transmission is feasible, but this approach is limited as the information provided by the separate layers is lost in the process. A DAC is fairly popular means of studying paint samples if available, particularly if the separate layers can be separated.

Forensic paint samples are commonly small chips so infrared microscopy is a valuable tool, especially for differentiating the composition of the various layers. Microscopy can be used to record both transmission and reflectance spectra. Cross-section samples of paint are prepared by embedding in a suitable media (e.g. acrylic, polyester or epoxy resin) and microtoming with cutting devices made of diamond, steel, tungsten carbide or glass to produce very thin sections. To produce a good-quality spectrum, a paint sample needs to be of the order of 1–10 μm in thickness. FTIR imaging is emerging as a new analytical approach to

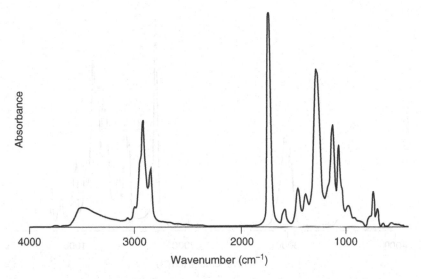

Figure 4.3 Infrared spectrum of an alkyd resin.

the study of paint cross-sections. This technique provides information about the spatial distribution of the characteristic functional groups in a paint cross-section. Infrared imaging has been more widely applied to the study of artworks, but, with improved accessibility, will find increasing use in forensic science applications.

Infrared spectroscopy is a particularly good technique for identifying the paint binder. The infrared spectra of some common binders encountered in forensic paint samples (alkyd resin, acrylic resin and PVA) are shown in Figures 4.3–4.5. The technique is also helpful for identifying the type of pigment or extender used.

SAQ 4.2

Figure 4.6 shows the micro-ATR spectrum obtained for an unknown blue paint specimen. Identify the binder used in the paint. Is it possible to identify the pigment in this paint using the infrared spectrum?

4.2.4 Fibres

Infrared spectroscopy may be used to characterize both synthetic and natural fibres [1, 3, 10–12]. There is a choice of sampling techniques. If a fibre is not too thick, a transmission approach may be taken. The fibre can be taped at the ends across a hole in a metal disk or by simply laying on an infrared window. However, in general fibres are too thick for transmission techniques,

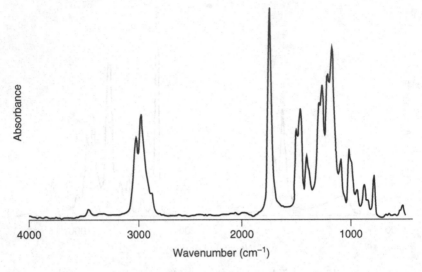

Figure 4.4 Infrared spectrum of poly(methyl methacrylate).

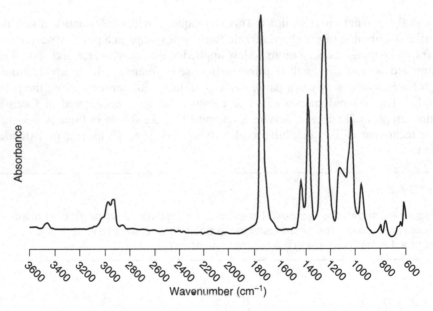

Figure 4.5 Infrared spectrum of poly(vinyl acetate).

with diameters less than 20 μm required to obtain good-quality spectra. Fibres can be flattened to reduce the thickness and increase the surface area. A DAC is a useful device for flattening fibres. Infrared microscopy is the most useful technique for obtaining fibre spectra. Infrared microscopy can be used to produce transmission or reflectance spectra, where thicker fibres can be examined without modification.

Given the variety in composition of synthetic fibres, infrared spectral databases containing the spectra of different classes and sub-classes of polymer fibres have proved useful. The infrared spectra of some common polymer components used to produce fibres are illustrated in Figures 4.7–4.15. The ability to identify the subclass of synthetic fibres arises from the sensitivity of infrared spectroscopy to relatively subtle changes in structure. An example is provided by a comparison of the spectra obtained for nylon 6 and nylon 12 (Figures 4.14 and 4.15). A clear difference is observed in the relative intensities of the C–H stretching band in the $3000 - 2800 \, cm^{-1}$ region. The more intense band for nylon 12 results from the longer methylene chain in the nylon 12 molecule compared to that of nylon 6.

Apart from different polymers used within fibre subclasses, variation in commercial fibres exists due to compositional differences, an example being acrylic fibres. Most acrylic fibres are a copolymer of acrylonitrile with methyl methacrylate, methyl acrylate or vinyl acetate. Figure 4.16 shows the infrared spectra of acrylonitrile and acrylonitrile–methyl methacrylate fibres. A useful approach to

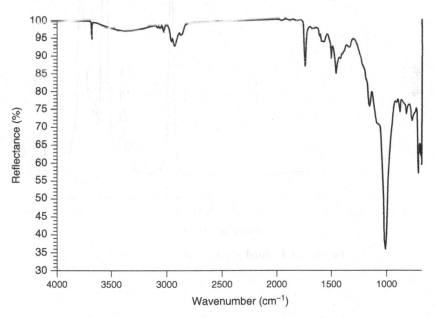

Figure 4.6 ATR infrared spectrum of a blue paint (cf. SAQ 4.2).

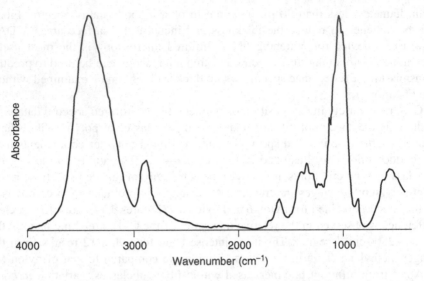

Figure 4.7 Infrared spectrum of regenerated cellulose.

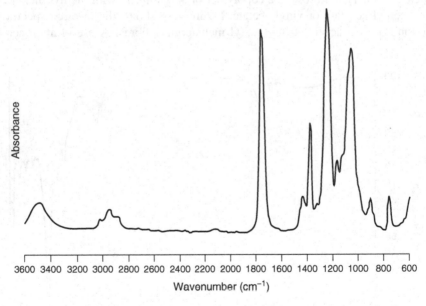

Figure 4.8 Infrared spectrum of cellulose acetate.

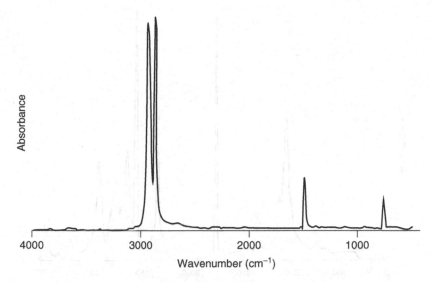

Figure 4.9 Infrared spectrum of polyethylene.

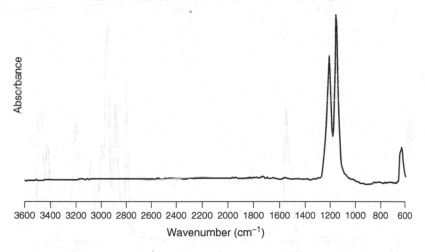

Figure 4.10 Infrared spectrum of polytetrafluoroethylene.

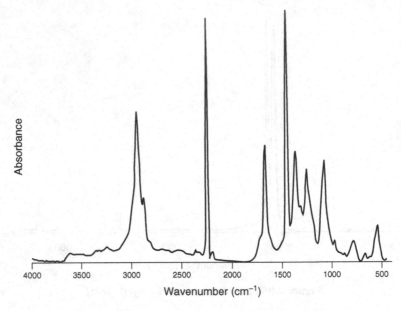

Figure 4.11 Infrared spectrum of polyacrylonitrile.

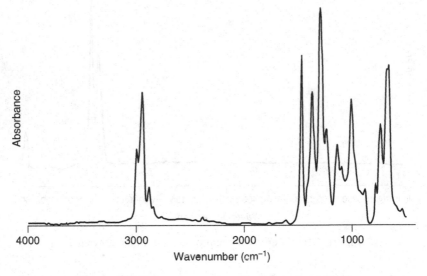

Figure 4.12 Infrared spectrum of poly(vinyl chloride).

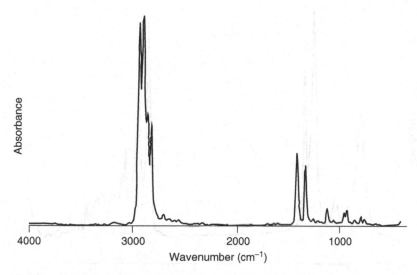

Figure 4.13 Infrared spectrum of polypropylene.

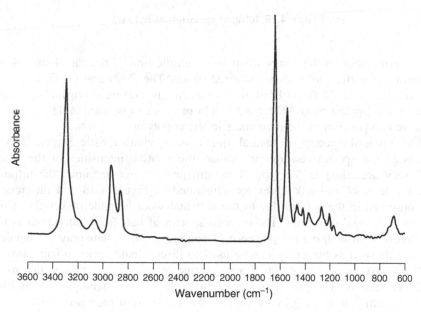

Figure 4.14 Infrared spectrum of nylon 6.

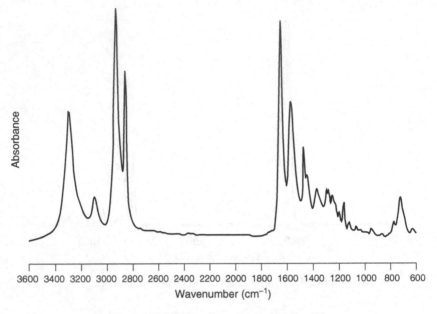

Figure 4.15 Infrared spectrum of nylon 12.

the determination of the composition of an acrylic fibre is the calculation of the ratio of the nitrile and carbonyl infrared bands. The C≡N and C=O stretching bands appear near 2240 and 1730 cm^{-1}, respectively, and are strong and relatively free of overlapping bands. The band height or area can be used to determine the relative composition of the monomers in the copolymer.

The infrared spectra of natural fibres show characteristic protein bands. Although the spectra can appear similar, the relative intensities of the bands will vary according to the type of protein present. For example, the infrared spectra of wool and silk fibres are illustrated in Figure 4.18 and differences are observed in the spectra due to the fact that wool is made of keratin, while silk is composed of fibroin. The infrared spectra of hair from individuals is not easily distinguished, but the presence of coatings such as hairspray or chemical treatments such as bleaching can be used to discriminate between hair samples [13]. The oxidation of the amino acid cystine in hair to cysteic acid due to an oxidizer such as hydrogen peroxide results in the S=O stretching bands at 1040 and 1175 cm^{-1} increasing in intensity in the spectrum of bleached hair.

4.2.5 Polymers

Infrared spectroscopy is a valuable technique for examining many types of polymer-based forensic evidence [1, 3, 13]. Packaging materials can be readily identified – the infrared spectra of common packaging materials PE, PP and PS

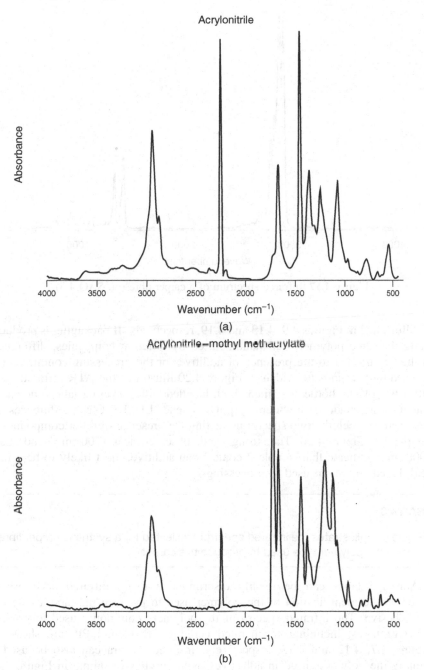

Figure 4.16 Infrared spectra of acrylonitrile and acrylonitrile – methyl methacrylate fibres.

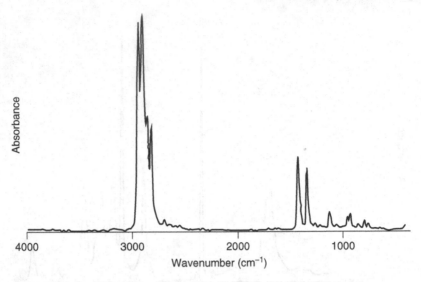

Figure 4.17 Infrared spectrum of a carpet fibre (cf. SAQ 4.3).

are illustrated in Figures 4.9, 4.13 and 4.19, respectively. If packaging is produced using the same polymer, but manufactured by different companies, differences in the spectra due to the presence of additives or the processing conditions can be used to discriminate evidence. Figure 4.20 illustrates the ATR infrared spectra of two plastic bottle specimens both labelled with recycling labels indicating that they are made of low-density polyethylene (LDPE). Clear differences are observed between the two spectra indicating the presence of other compounds in sample 2 in Figure 4.20. The strong bands observed near $1700\,\mathrm{cm}^{-1}$ and above $3000\,\mathrm{cm}^{-1}$ indicate that sample 2 contains an additive, most likely to be a fatty acid–based slip agent used in processing.

SAQ 4.3

Figure 4.17 illustrates the infrared spectrum collected for a synthetic carpet fibre. Identify the polymer type used to produce this carpet.

Adhesive tapes can be readily discriminated using infrared spectroscopy based on the identification of the polymer used to produce the tape, as well as the adhesive. The infrared spectra of the polymers commonly used to produce adhesive tapes, including regenerated cellulose, PVC and PP, are shown in Figures 4.7, 4.12 and 4.13, respectively. Infrared spectra can also be used to identify the polymers used in adhesives, with acrylics (example in Figure 4.4), styrene (Figure 4.19), isoprene (Figure 4.21) or butadiene (Figure 4.22) being possible components to be identified.

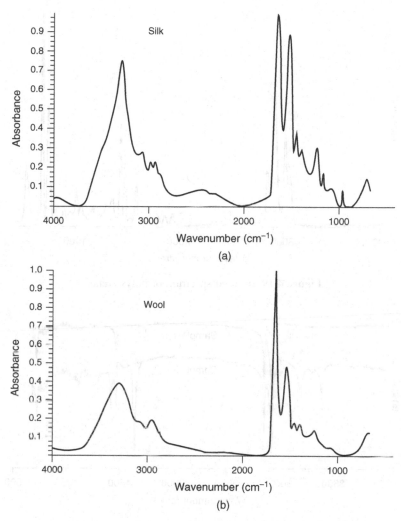

Figure 4.18 Infrared spectra of wool and silk fibres.

4.2.6 Documents

Different aspects of questioned documents can be investigated using infrared spectroscopy, including toners, inks and paper [2, 3, 13, 14]. ATR techniques are particularly applicable as they avoid the need for destructive preparative methods required in other techniques. For toners and inks, it is possible to differentiate the organic binders present. For example, the ability to identify differences in polymer type, copolymer composition and the presence of additives enables infrared spectroscopy to be used to classify toners.

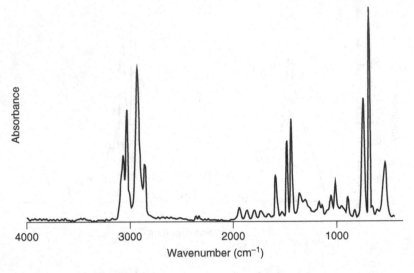

Figure 4.19 Infrared spectrum of polystyrene.

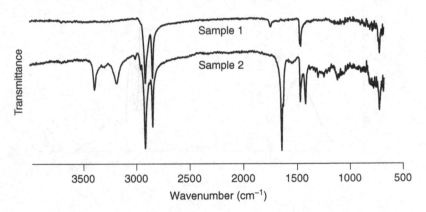

Figure 4.20 Infrared spectra of polyethylene packaging.

4.2.7 *Explosives*

FTIR spectroscopy can be used to identify explosive materials [3, 15–17]. A range of sampling techniques may be utilized to examine explosives given the variety of specimen types. Solids can be studied using KBr discs or mulls, and a liquid cell can be used for solutions or liquid explosives. ATR has emerged as a popular approach, especially ATR microspectroscopy. Portable FTIR spectroscopy lends itself to the field analysis of explosives.

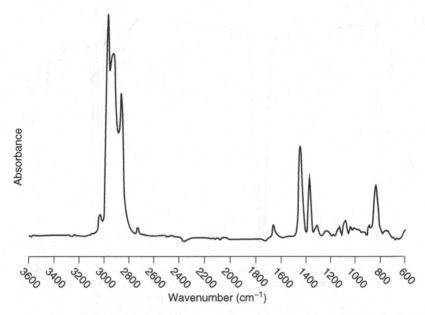

Figure 4.21 Infrared spectrum of poly(*cis*-isoprene).

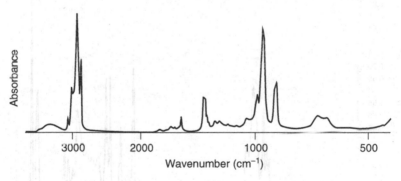

Figure 4.22 Infrared spectrum of polybutadiene.

4.2.8 Drugs

Infrared spectroscopy can be used to discriminate drug components that might otherwise be indistinguishable using other techniques [1, 3]. Drug samples can be examined using KBr discs or the ATR of individual particles. Samples from clandestine laboratories can be characterized using infrared spectroscopy. For

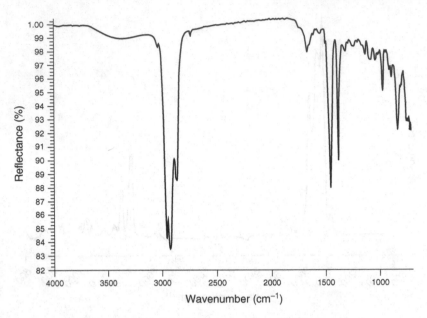

Figure 4.23 Infrared spectrum of an adhesive tape.

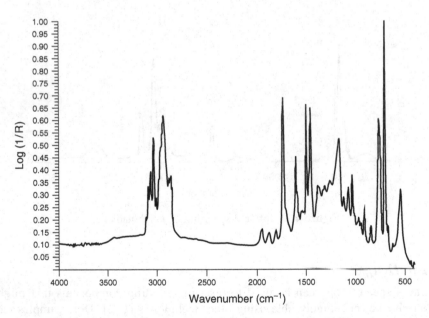

Figure 4.24 Diffuse reflectance infrared spectrum of a black toner.

example, the two amphetamines methamphetamine and phenteramine have very similar molecular compositions and it is difficult to distinguish these structures using other techniques, but their infrared spectra show useful differences. In addition to the quite different patterns observed in the fingerprint region, there are observable differences in the C–H stretching region near $3000\,cm^{-1}$ due to the different CH_3 stretching of the methyl groups adjacent to the primary and secondary amines in the respective molecules.

SAQ 4.4

The micro-ATR spectrum of the adhesive side of an adhesive tape is shown in Figure 4.23. Identify the type of adhesive type used to produce the tape.

SAQ 4.5

Figure 4.24 shows the diffuse reflectance infrared spectrum obtained for a black toner sample. Identify the type of resin present in this toner.

SAQ 4.6

What are the differences in the infrared spectra of pseudoephedrine and methamphetamine that may be used to discriminate these compounds?

4.3 Raman Spectroscopy

Raman spectroscopy is a technique that involves the study of how radiation is scattered by a sample [1, 3, 18]. Most scattered radiation is unchanged in wavelength and is known as Rayleigh scattering; a small amount of the scattered light is slightly increased or decreased in wavelength, and this is known as Raman scattering. When the wavelength is increased, the process is known as Stokes Raman scattering, while a decreased wavelength is associated with anti-Stokes Raman scattering. Raman scattering by molecules involves transitions between rotational or vibrational states. In order to be Raman active, a molecular rotation or vibration must cause a change in a component of the molecular polarizability – a measure of the degree to which a molecule can have its positive and negative changes separated.

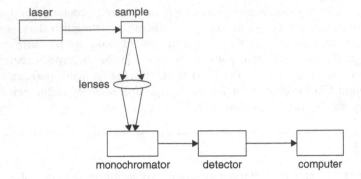

Figure 4.25 Layout of a Raman spectrometer.

4.3.1 Methods

The typical layout of a Raman spectrometer is shown in Figure 4.25. The source of radiation used in a Raman spectrometer may be in the near-ultraviolet, visible or near-infrared regions of the spectrum. He–Ne lasers at a wavelength of 632.8 nm or an Ar laser at 514.5 nm are often used as an excitation source. A consideration when choosing a source is the potential for fluorescence – this can be produced by the sample or impurities within the sample and can swamp the spectrum. However, fluorescence may be minimized by using a source of higher wavelength near-infrared sources, and a Nd–YAG laser at 1064 nm is used in a Fourier transform Raman spectrometer (where weaker spectra can be collected). The scattered light from a sample passes through a series of focussing and collection optics. Optical filters are used for the rejection of Rayleigh scattered light, and the light is sent to the detector.

Raman microscopy is a valuable forensic tool and enables spectra with μm spatial resolution of samples to be investigated. A typical set-up involves combining a microscope with a spectrometer. The sample is placed on the stage of the microscope, illuminated by white light and brought into focus by adjusting an objective. The illuminator lamp is then switched off and the laser radiation is directed to the beamsplitter. The scattered light from the sample is collected by the objective and is sent to the spectrometer.

The development of portable and more cost-effective instruments has led to wider adoption of the technique for forensic applications. Handheld Raman instruments have the same basic design as laboratory-based instruments, but are more compact in design. A 532 nm laser is commonly used in portable instruments. Fibre-optic sampling is also feasible with Raman spectroscopy as visible or near-infrared radiation can be transmitted over considerable distances through optical fibres, enabling samples to be examined *in situ* using portable Raman spectrometers.

Several resonance Raman techniques are also available to enhance the intensity of Raman spectra. *Resonance Raman spectroscopy* (RRS) uses incident radiation that nearly coincides with the frequency of an electronic transition of the sample. During the process, an electron is promoted into an excited electronic state followed by an immediate relaxation to a vibrational level of the electronic ground state. *Surface-enhanced Raman spectroscopy* (SERS) involves obtaining the Raman spectra of samples adsorbed on metal surfaces, such as silver or gold, and the intensity of certain bands in the spectrum is enhanced for a molecule on a metal.

Raman spectroscopy has the advantage that a variety of sample types can be examined with minimal preparation. Solids in the form of powders, films or fibres can be studied, and liquids can be examined using capillaries or ampules. If there is a risk of overheating the sample with the laser, the sample may be cooled or a rotating cell might be used. Multi-layered samples, such as paint chips, can be examined by embedding in a resin, sectioned and mounted so that the cross-section of each of the layers may be examined.

4.3.2 Interpretation

Raman spectra are plotted as intensity as a function of the wavenumber shift from the excitation radiation, $\Delta\bar{\nu}$. However, it is common to see just wavenumber used on the scale as Stokes Raman bands are shown.

Raman spectroscopy provides a complementary technique to infrared spectroscopy. They are both forms of vibrational spectroscopy, but result from the different selection rules. Bands due to particular functional groups appear with different intensities in the two techniques. As with infrared spectroscopy, correlation tables may be consulted to identify bands appearing in a Raman spectrum. However, the most common approach to the characterization of forensic samples is by comparison with a spectral library or with a reference spectrum of a known source.

4.3.3 Drugs

Raman spectroscopy has recently been expanding in use for the identification of illicit drugs in powder, tablet or liquid form in the laboratory and in the field [1, 3, 18]. Drug molecules tend to be good Raman scatterers, and cutting agents may also be identified using this approach. The use of fibre optics enables illicit substances contained in plastic bags and bottles to be examined. An example of where Raman spectroscopy can be used to differentiate drug compounds of similar structure is a comparison of the spectra of amphetamine and methamphetamine. Figure 4.26 shows the Raman spectra of these compounds. and the two drugs can be discriminated using a band at 2459 cm^{-1}. This band is characteristic of a secondary amine and appears in the methamphetamine spectrum due to the methyl substituent of the amine.

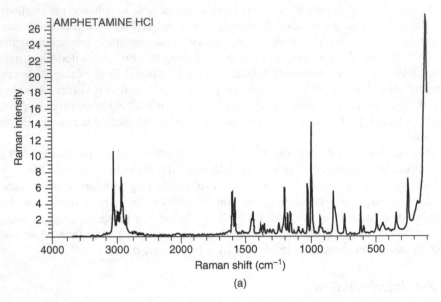

(a)

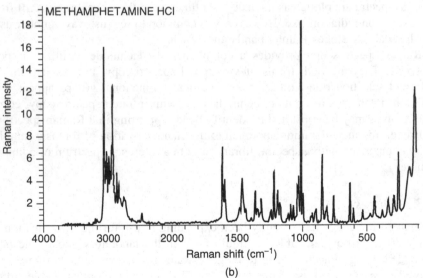

(b)

Figure 4.26 Raman spectra of amphetamine and methamphetamine. Reproduced with permission from T. Mills, J.C. Roberson, C.C. Matchett, M.J. Simon, M.D. Burns and R.J. Ollis (eds), *Instrumental Data for Drug Analysis*, 3rd ed., CRC Press, Boca Raton, 2005, Taylor & Francis.

SAQ 4.7

What would be the expected differences in the Raman spectra of heroin and morphine?

4.3.4 Paint

Raman spectroscopy is highly applicable to the forensic analysis of paints due to the ease of identification of the component pigments [1, 3, 18–20]. The technique is particularly good for identifying inorganic pigments, which produce good-quality Raman spectra showing sharp characteristic bands in the low-wavenumber region of the spectra due to lattice vibrations. Table 4.2 lists Raman bands observed for some common pigments and extenders [21]. Pigments can produce fluorescence that may mask the Raman spectra, but the use of longer excitation wavelength can avoid this problem. Raman spectroscopy has the added advantage that it may be used to differentiate pigments with the same chemical formula,

Table 4.2 Raman bands of some common pigments and extenders

Pigment or extender	Laser source (nm)	Raman band (cm^{-1})
anatase	785	640, 520, 400
	1064	144, 201, 397, 512, 634
calcite	1064	154, 282, 712, 1086
	514.5	157, 282, 1088
chrome yellow	785	845, 405, 365
gypsum	1064	140, 181, 493, 619, 670, 1007, 1132
	514.5	181, 414, 493, 619, 670, 1007, 1132
haematite	632.8	224, 243, 290, 408, 495, 609
	1064	224, 244, 292, 409, 610
lead molybdate orange	785	880, 825, 360, 345, 145
Prussian blue	785	2150, 2090, 275
	514.5	282, 538, 2102, 2154
rutile	785	1450, 1000, 980, 615, 450
	1064	144, 232, 447, 609, 147, 242, 440, 611

but of different crystalline structure (polymorphs). Paint binders and varnishes may also be identified using Raman spectroscopy, although these tend to be more successfully examined by infrared spectroscopy. Paint chips with multiple layers may be examined using Raman spectroscopy to characterize the composition of individual layers.

SAQ 4.8

Outline a suitable Raman spectroscopy experiment that could be used to determine what type of TiO_2 pigment has been used in a small paint chip.

4.3.5 Fibres

Raman spectroscopy can be used to examine textile fibres for forensic purposes [1, 3, 18, 22]. Raman microscopy is suitable for the examination of fibres and can be used to determine the composition of the fibres. The technique may also be used to discriminate coloured fibres: many dye chromophores produce resonance Raman effects, so even low concentrations of dye can be detected on textile fibres.

4.3.6 Documents

The nondestructive nature of Raman spectroscopy allows for the examination of questioned documents and is mainly based on the characterization of the inks used [1, 23]. The inks can be identified based on the varying composition of dyes, pigments or polymers. The sequence of handwritten lines may also be examined with Raman microscopy as it is possible to examine the surface layer on a paper surface using this technique. Thus, it is possible to determine the composition of the surface layer without interference from the layers beneath.

SAQ 4.9

A paper document is examined using Raman spectroscopy with a laser source of 514.5 nm. The resulting spectrum shows a large broad band with no discernable peaks. What is responsible for the appearance of the spectrum? How would you go about remedying this situation?

4.3.7 Explosives

Raman spectroscopy can be used to identify a range of common explosives and their precursors [1, 3, 16, 18, 24, 25]. It is possible to examine trace amounts of explosives on many surfaces such as clothing and painted or metallic surfaces. Samples can be examined in the laboratory using Raman microscopy or in the field using portable instruments. Table 4.3 lists the Raman bands for some common explosives and their precursor compounds.

Table 4.3 Major Raman bands of some common explosives and explosive precursors

Explosive or precursor	Wavenumber (cm^{-1})	Assignment
PETN	1290	symmetric NO_2 stretching
	871	symmetric O–N stretching
	622	C–C–C bending
pentaerythritol (precursor for PETN)	873, 810	CH_2 rocking
	439	C–C–C bending
TNT	1532	asymmetric NO_2 stretching
	1357	symmetric NO_2 stretching
	822	NO_2 scissoring
ammonium nitrate	712	NO_3^- bending
	462, 1040	N–C–N bending
RDX	1210–1310	symmetric NO_2 stretching
	1340–1440	CH_2 skeletal bands
hexamethylenetetramine (precursor for RDX)	777	C–N stretching
	1071	C–O stretching

SAQ 4.10

Figure 4.27 illustrates the FT-Raman spectra of RDX, PETN and Semtex. Use the spectra to identify the main explosive present in the Semtex sample.

4.4 Ultraviolet–visible Spectroscopy

Ultraviolet–visible (UV–vis) spectroscopy involves the examination of the electronic transitions associated with absorptions in the UV (180–390 nm) and visible (390–780 nm) regions of the electromagnetic spectrum [26]. The energies associated with these regions are capable of promoting the outer electrons of a molecule from one electronic energy level to a higher level. The part of the molecule containing the electrons involved in the electronic transition responsible for the observed absorptions is called the chromophore. The types of transitions that result in UV–vis absorptions consist of the excitation of an electron from the highest occupied molecular orbital (usually of nonbonding *p* or bonding π orbital) to the next lowest unoccupied molecular orbital (an anti-bonding $\pi*$ or $\sigma*$ orbital). Nonbonding orbitals are represented by *n*, and an asterisk is used to represent an anti-bonding orbital.

4.4.1 Methods

A UV–vis spectrophotometer consists of a UV–vis radiation source, a sample compartment, a dispersing element and a detector [26]. Often there are two light

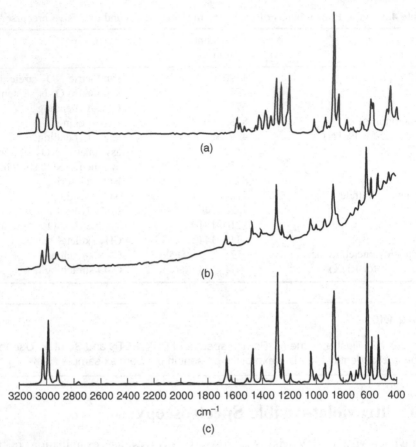

(a)

(b)

3200 3000 2800 2600 2400 2200 2000 1800 1600 1400 1200 1000 1800 600 400

cm⁻¹

(c)

Figure 4.27 Raman spectra of: (a) RDX; (b) Semtex; (c) PETN of (SAQ 4.10). Reproduced from J. Akhavan, *Spectrochim. Acta* A **47**,1247–1250, 9–10, with permission from Elsevier Science (1991).

sources; a deuterium lamp for UV light and a tungsten lamp for visible light. Single-beam UV–vis spectrometers are set up so that the spectrum of the reference solution is measured first, followed by that of the sample of interest. Double-beam spectrometers separate the light into two parallel beams passing through two different cells, with one cell containing the reference solvent and the other containing the sample of interest. Cells for recording the UV–vis spectra of solutions and liquids are commonly made of quartz or glass (usually 1 cm pathlength).

As forensic samples can be limited in quantity, the ability to examine microscopic specimens is important and this is achieved by using UV–vis *microspectrophotometry* (MSP) [27]. MSP functions in a similar manner to infrared and Raman microscopy, that is, a spectrometer that detects in the

UV–vis range is combined with optical microscopy. Visible microspectrometers allow objects smaller than 10 μm in size to be nondestructively analysed. Samples may be examined in reflectance mode, and a spatial resolution of about 1 μm can be achieved.

4.4.2 Interpretation

Generally dilute solutions are examined in UV–vis spectroscopy and the intensity of the transmitted light is related to the concentration of the absorbing molecule by Beer's law. The law shows that the absorbance (A) of a solution is proportional to the concentration of the absorbing molecule (c) and the pathlength of the cell (l) containing the solution:

$$A = \varepsilon cl \tag{4.1}$$

The constant ε is the molar absorptivity of the absorbing molecule. Thus, a plot of absorbance versus concentration will be linear with a gradient of εl and will pass through the origin. The higher the ε value, the greater the probability of the electronic transition. The absorbance is determined by measuring the ratio of the intensity of the incident light (I_0) and the intensity of light transmitted through the sample (I):

$$A = \log_{10} (I_0 / I) \tag{4.2}$$

The bands in the UV–vis spectrum are usually quite broad, and the wavelength of the maximum (λ_{max}) is assigned. Derivative spectroscopy allows the resolution of UV–vis data to be enhanced. Likewise, chemometric techniques such as PCA are used to gather information from apparently similar spectra.

UV–vis spectra are sensitive to differences in solvent, pH and conjugation. Changing the solvent results in a change in the difference between the electronic energy levels of a molecule and, hence, a shift in the wavelength of the associated absorption band. The more conjugation appearing in a molecule, the greater the intensity of the absorption bands of the molecule and the transitions appear at higher wavelengths. Likewise, the pH of the solvent affects the spectra as the addition or removal of photons in a molecule changes the observed electronic transitions. When there is a shift in the λ_{max} towards longer wavelengths, it is known as a red or bathochromic shift. When the shift is towards a shorter wavelength, the change is known as a blue or hypsochromic shift.

4.4.3 Fibres

As the analysis of the colour of a fibre may be critical to the discrimination of such evidence, UV–vis spectra can prove valuable [27–30]. UV–vis spectroscopy produces the characteristic spectra of pigments. Although two fibres may appear identical in colour to the human eye, the fibres may have been dyed with different pigments and visible spectroscopy is capable of separating the different compounds used in manufacture. Embedding solvents, such as glycerine, are used to fix the sample fibres and must not produce absorbance or fluorescence in the spectral region studied. Because of the size of fibre evidence, conventional UV–vis spectroscopy is not an appropriate choice, but MSP is very suitable.

Organic pigments and dyes can show transitions between delocalized molecular orbitals. The bands produced are very intense and usually occur in the visible or very near UV region. Charge transfer transitions involve transitions between molecular orbitals at different sites of a molecule where there is electron transfer. Such transitions are observed for certain pigments. Many inorganic pigments show ligand field transitions, which occur between energy levels that are localized mainly on a metal ion. Table 4.4 summarizes the regions of the spectrum where different colours are observed [26].

UV–vis bands are quite broad, so the use of derivative spectra is helpful for enhancing differences in the spectra. Figure 4.28 illustrates the absorbance spectrum of a red acrylic fibre recorded using visible MSP and shows a broad band with a maximum near 535 nm and a subtle shoulder near 450 nm. The first derivative of this absorbance spectrum is also shown in Figure 4.28 and demonstrates that such spectral features are more distinctive when a derivative is used.

4.4.4 Paint

MSP enables the colours of paint specimens to be characterized and, as with fibres, colours that appear similar to the eye can be readily discriminated [8, 27, 29, 31]. Reflectance spectra can be obtained from the surface layers of paint or the individual layers observed in an embedded cross-section. The technique allows the colour coordinates to be determined. The tristimulus values are mathematical representations of colour and the differences between colours are represented by

Table 4.4 Colours detected in the visible spectrum

Wavelength (nm)	Observed colour
390–420	green-yellow
420–440	yellow
440–470	orange
470–500	red
500–520	purple
520–550	violet
550–580	violet-blue
580–620	blue
620–680	blue-green
680–780	green

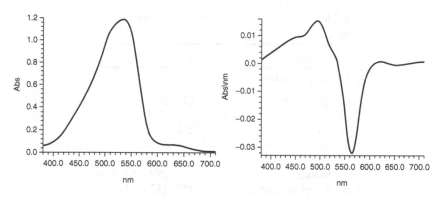

Figure 4.28 UV-visible absorbance (left) and first derivative (right) spectra of a red acrylic fibre. Reprinted from K. Wiggins et al., *Science and Justice* **47**, 9–18 (2007) with permission from Elsevier.

the Cartesian distances between the colours on a plot. Transmittance UV–vis spectra can also be used to identify additives, enabling further discrimination of paint samples.

4.4.5 Documents

The visible spectra of inks provide a means of discriminating similarly coloured inks [27, 28]. Inks are commonly extracted from paper using an ethanol–water mixture. As the visible spectra of inks tend to be broad in appearance, the use of chemometrics is helpful and PCA is a recommended approach. The UV–vis spectrum of black ballpoint pen ink in the 300–800 nm region provides a means of identifying inks from different sources based on the dyes present.

4.4.6 Drugs

Drug molecules produce characteristic UV–vis spectra, and generally the spectra are commonly recorded at low and high pH values and then compared with known standards [32]. However, UV–vis spectroscopy is less useful as a technique for confirming the identification of a drug due to the nature of the spectra produced, as UV–vis spectra show broad overlapping bands that are less specific than, say, infrared or Raman bands. Another problem is that the presence of another drug or substance may interfere with the analysis. However, if a sample is known to contain a single absorber, it is possible to carry out quantitative analysis. Table 4.5 lists the characteristic UV maxima for some commonly encountered drug molecules recorded in both acidic and basic solutions [32].

Table 4.5 UV absorption maxima for common drugs

Compound	λ (nm)
In acidic solution	
amphetamine	251, 257[a], 263
cocaine	233[a], 275
codeine	284
heroin	278
LSD	224[a], 310
methadone	253, 259, 264, 292[a]
methamphetamine	252, 257[a], 263
pseudoephedrine	251, 256[a], 262
In basic solution	
amphetamine	258[a], 267
cocaine	230[a], 273
codeine	284
LSD	239[a], 308
methamphetamine	258[a], 267
morphine	298
pseudoephedrine	251, 257[a], 263

[a] Major peak.

SAQ 4.11

Is it possible to use UV–vis spectroscopy to differentiate samples of amphetamine and methamphetamine?

SAQ 4.12

A sample of cocaine dissolved in water is to be quantitatively analysed using UV–vis spectroscopy. A series of standard solutions of cocaine in water are

Table 4.6 Calibration data for cocaine solutions (cf. SAQ 4.12)

Concentration (μg ml^{-1})	Absorbance at 233 nm
30.0	0.838
25.0	0.699
20.0	0.559
15.0	0.419
10.0	0.279

prepared, and the spectrum is recorded for each over the range of 210–350 nm. The concentrations and absorbance readings obtained for each at $\lambda_{max} = 233$ nm are summarized in Table 4.6. A sample of cocaine is also dissolved in water and the spectrum recorded. The absorbance reading at 233 nm is found to be 0.432. Determine the concentration of cocaine in the unknown solution. What assumption has been made for the calculation of the cocaine concentration?

4.4.7 Toxicology

UV–vis spectroscopy can be used for quantitative analysis in the field of forensic toxicology. Following appropriate sample pre-treatment steps, such as filtration, centrifugation or pH adjustments, the spectra of toxicological samples can be recorded and the absorbance produced by the analyte of interest can be measured. An example of the approach is the determination of the CO concentration in blood. A common approach is to measure the concentration of COHb using a UV–vis spectrophotometer [33]. The spectrum of a blood sample with sodium dithionite ($Na_2S_2O_4$) added is recorded. The addition of $Na_2S_2O_4$ reduces the oxyhaemoglobin (the normal oxygenated haemoglobin complex) while leaving the COHb unchanged. The CO level is calculated from the spectrum by measurement of the ratio of the absorbance values at 541 and 555 nm. A calibration plot can be established by measuring this ratio from a series of solutions of known CO concentration.

4.5 Fluorescence Spectroscopy

The phenomenon of fluorescence was introduced in Chapter 2, where its use in visualization was explored. Fluorescence is also used in forensic science in the form of a spectroscopic technique known as fluorescence spectroscopy, fluorimetry or spectrofluorimetry [34]. During fluorescence, molecular collisions cause the electronically excited molecule to lose vibrational energy until it reaches the lowest vibrational energy level of the electronically excited state. From here, the molecule drops to a vibrational energy level in the ground electronic state and a photon is emitted during the process. The types of molecules that can be examined using this approach are particular aromatic or conjugated molecules.

4.5.1 Methods

In a fluorimeter, fluorescence spectra are measured by viewing the emission at angles of 90° to the direction of the excitation (Figure 4.29). The exciting light is selected in a monochromator, and a second monochromator is then used to scan the emitted light from near the excitation wavelength up to longer wavelengths.

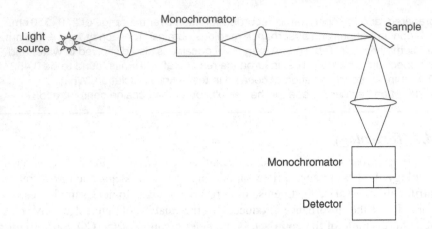

Figure 4.29 Layout of a fluorescence spectrometer.

The ability to determine the often small forensic specimens is enabled by the use of a microspectrofluorimeter. Solid and liquid samples can be examined. As fluorescence intensity depends on the excitation wavelength, the latter is varied to obtain the most suitable value for the sample in question.

4.5.2 Interpretation

An emission spectrum is a plot of the intensity of light emitted as a function of wavelength. The spectrum generally shows broad, seemingly featureless bands. The wavelength of maximum intensity is often used for identification purposes. Given the nature of the spectra, multivariate statistical analysis is a valuable analytical approach. Like UV–vis spectroscopy, Beer's law can be used in fluorescence spectroscopy. As with UV–vis spectroscopy, a set of standards of known concentrations can be used for quantitative analysis. Thus, the concentration of analytes can be determined from a fluorescence spectrum. Although only a relatively limited number of molecules fluoresce, fluorescence spectroscopy is a sensitive technique because small amounts of fluorescing materials can be detected and quantitatively analysed.

4.5.3 Body Fluids

Fluorescence spectroscopy is a sensitive technique for the detection of body fluids [35, 36]. Many of the components of body fluids exhibit fluorescence. As the composition of each fluid is different, characteristic emission spectra can be recorded. An advantage of using fluorescence spectroscopy is that no destructive reagents are used in the technique. Although specimens can be examined over a range of excitation wavelengths to aid in the identification of a body fluid, care

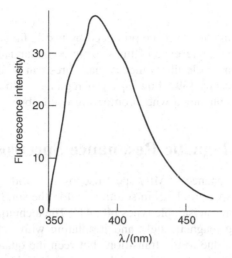

Figure 4.30 Fluorescence spectrum of LSD in ethanol.

must be taken in the choice of wavelength because certain energies can cause degradation.

4.5.4 Toxicology

Particular drugs exhibit fluorescence and may be studied using fluorescence spectroscopy. A common example is LSD, which strongly absorbs light at 320 nm and emits light near 400 nm. Figure 4.30 illustrates the fluorescence spectrum of LSD in ethanol. Drugs in body fluids such as urine or blood can also be effectively analysed using this approach. For example, quinine is a common adulterant of illicit heroin samples and fluorescence spectroscopy can be used to identify the presence of quinine as a screening test for heroin [37, 38]. The quinine can be extracted to produce an aqueous solution suitable for fluorimetric analysis.

SAQ 4.13

A urine sample collected for drug analysis is tested for quinine. Quinine is extracted with chloroform and re-extracted with sulfuric acid. A fluorimeter with an excitation wavelength of 350 nm is used to record an emission spectrum for the sample at 445 nm. A 1 mg l^{-1} quinine sulfate stock solution is prepared, and the fluorescence spectrum is recorded under the same conditions. If the intensity of the quinine fluorescence in the urine extract is 0.76 and that of the quinine standard is 0.92, determine the concentration of quinine in the urine sample. Suggest an improvement to this method to ensure an accurate measurement of the quinine concentration.

4.5.5 Fibres

The examination of the fluorescence produced by textile fibres was introduced in Chapter 2. The common presence of fluorescing compounds including dyes and optical brighteners in textile fibres means that fibres can be discriminated using fluorescence spectroscopy [39]. Four spectral regions – ultraviolet, violet, blue and green – are recommended when comparing dyed fibres.

4.6 Nuclear Magnetic Resonance Spectroscopy

Nuclear magnetic resonance (NMR) spectroscopy is a widely used tool for the study of molecular structure, both in solution and in the solid state, and is applicable to particular forensic sample types [40, 41]. The technique involves placing a sample in a strong magnetic field and irradiating with radiofrequency radiation. The absorptions due to the transitions between the quantized energy states of nuclei oriented by the magnetic field are observed. Nuclei carry a charge and, in particular nuclei, the charge spins on the nuclear axis, generating a magnetic dipole. The angular momentum of the spinning charge is described using a spin number (I). Each proton and neutron has its own spin, and I is the result of these spins. NMR spectroscopy utilizes nuclei with a $I = \frac{1}{2}$ such as H^1, C^{13}, Al^{27}, Si^{29} and P^{31}, and for such nuclei, the magnetic moments align either parallel to or against the magnetic field. The energy difference between the two states is dependent upon the magnetogyric ratio (γ) of the nucleus and the strength (B) of the external magnetic field. The resonance frequency (ν), which is the frequency of radiation required to effect a transition between the energy states, is given by:

$$\nu = \frac{\gamma B}{2\pi} \tag{4.3}$$

4.6.1 Methods

Figure 4.31 illustrates the layout of a NMR spectrometer. NMR spectrometers possess a strong magnet with a homogenous field, a radiofrequency transmitter, a radiofrequency receiver and a recorder. The instrument contains a sample holder, which spins the sample to increase the apparent homogeneity of the magnetic field. Modern NMR spectrometers use superconducting magnets (e.g. 500 MHz or 800 MHz), and a short pulse of radiofrequencies is applied to promote the nuclei to higher energy states. The relaxation of the nuclei back to the lower energy state is detected as an interferogram known as free induction decay, which can be converted into a spectrum by Fourier transformation. Samples can be studied in solution (μg ml^{-1} quantities) and in solid form.

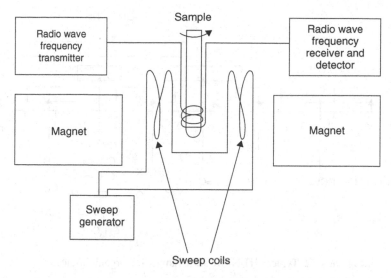

Figure 4.31 Layout of NMR spectrometer.

4.6.2 Interpretation

A single peak is theoretically predicted to result from the interaction of radiofrequency radiation with a magnetic field on a nucleus, but in practice absorptions at different positions are observed. This occurs because the nucleus is shielded to a small extent by its electron cloud, the density of which varies with the environment. The difference in the absorption position from that of a reference proton is known as the chemical shift (δ):

$$\delta = \frac{\text{frequency of absorption} \times 10^6}{\text{applied frequency}} \tag{4.4}$$

A common reference compound is tetramethylsilane (TMS). TMS produces ^{1}H and ^{13}C absorptions corresponding to $\delta = 0$. Each absorption area in a NMR spectrum is proportional to the number of nuclei it represents, and these areas are evaluated by integration. The ratios of integrations of absorptions are equal to the ratios of the number of the nuclei present in the nucleus. To aid in the interpretation of NMR spectra, extensive correlation tables of chemical shifts and libraries of spectra are available. Figure 4.32 illustrates the chemical shift regions in which various proton resonances are observed.

Quantitative NMR analysis can be carried out through the integration of peaks. For instance, the integrated area of a proton resonance peak of an internal standard

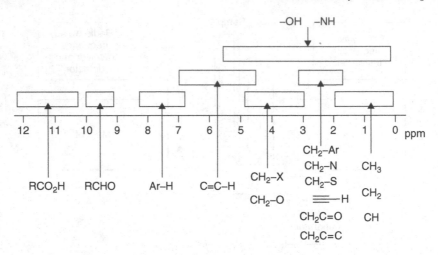

Figure 4.32 Typical ^{1}H NMR shift ranges for organic molecules.

is compared with the area of the resonance peak of the sample being analysed. The composition may be determined using:

$$\% \text{ composition} = \left(W_I / W_S\right) \times \left(A_S / A_I\right) \times \left(M_S / M_I\right) \times 100 \qquad (4.5)$$

where W_S and W_I are the weight of the total sample and the internal standard, respectively; A_S and A_I are the integrated areas of the sample and standard, respectively, and M_S and M_I are the proton equivalent weights (molecular weight of compound divided by the number of protons producing the absorption signal) of the sample and standard, respectively.

After the absorption of energy by the nuclei during an NMR experiment, there must be a mechanism by which the nuclei can dissipate energy and return to the lower energy state. Spin–lattice relaxation involves the transfer of energy from the nuclei to the molecular lattice. The spin–lattice relaxation time (T_1) is the time constant for the exponential return of the population of spin states to their equilibrium population. Spin–spin relaxation occurs from direct interactions between the spins of different nuclei that can cause relaxation without any energy transfer to the lattice. The spin–spin relaxation time (T_2) is the time constant for the exponential transfer of energy from one high-energy nucleus to another.

Spin–spin coupling complicates NMR spectra and is the indirect coupling of nuclei spins through intervening bonding electrons. This occurs because there is a tendency for a bonding electron to pair its spin with the spin of the nearest $I = \frac{1}{2}$ nuclei. The spin of the bonding electron influenced will affect the spin of the other bonding electron and so on through to the next $I = \frac{1}{2}$ nucleus. Where spin–spin coupling occurs, each nucleus will give rise to separated absorptions appearing as doublets as the spin of each nucleus is affected by the orientations of the other nucleus through the intervening electrons. The frequency difference between the doublet peaks is proportional to the effectiveness of the coupling and is known as a coupling constant (J).

The complexity of NMR spectra is also a result of the Nuclear Overhauser Effect (NOE), which produces a change in signal intensity of one spin due to the relaxation of a saturated spin that is coupled to the first spin. NOEs occur through space rather than through a bond interaction and so provide information about the distance between spins. Complex spectra can also be interpreted using two-dimensional (2D) spectra, where the data are plotted with two frequency axes to illustrate the interactions.

4.6.3 Drugs

NMR spectroscopy lends itself to the characterization of drug molecules as it is an established technique for the elucidation of chemical structures and is widely used for the identification of organic compounds. Drug analysis using NMR spectroscopy might involve the identification of the major component of a drug seizure or may involve the determination of minor components such as impurities, precursors and intermediates, as a means of discriminating samples [40–42].

Drug samples can be examined in either powder or liquid forms. It is possible to use a solvent that preferentially dissolves the drug of interest while not dissolving the other components such as sugars. The adjustment of pH can be used to further separate the drug components. If the drug under investigation is contained in a biological fluid, then it may be directly analysed or by analysis in deuterium oxide.

Both ^1H and ^{13}C NMR spectra of a broad range of drugs have been collected, and commercial spectral libraries are available to identify drug samples [32]. If the structures under investigation do not provide a straightforward match to library spectra, there are a number of data analysis techniques than can be used to elucidate structures, such as NOE or 2D experiments. With the advent of designer drugs, it is particularly important to be able to identify the exact position of a substituent group in a drug molecule.

SAQ 4.14

Figure 4.33 illustrates the ^1H NMR spectrum of amphetamine in deuterated chloroform. Use the correlation table shown in Figure 4.32 to assign the peaks in the spectrum.

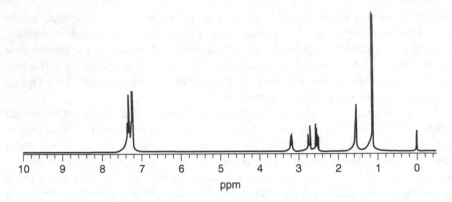

Figure 4.33 ¹H NMR spectrum of amphetamine in deuterated chloroform. Reproduced from T. Mills, J.C. Roberson, C.C. Matchett, M.J. Simon, M.D. Burns and R.J. Ollis (eds), Instrumental Data for Drug Analysis, 3rd ed., CRC Press, Boca Raton, with permission from Taylor & Francis, 2005.

Quantitative analysis of drug samples can be carried out using NMR spectroscopy. A common choice for an internal standard for drug analysis is maleic acid. Maleic acid shows a resonance peak due to two protons appearing at 6.2 ppm. Most drugs are free of resonances at 6–7 ppm.

SAQ 4.15

The NMR spectrum of the barbiturate phenobarbital in deuterium oxide shows a triplet at 1.0 ppm with a proton density of 3, a quartet at 2.5 ppm with a proton density of 2 and a resonance peak at 7.4 ppm with a proton density of 5. Using Figure 1.10, assign these peaks to the relevant phenobarbital structural components. Is maleic acid a suitable internal standard if quantitative analysis of phenobarbital is to be carried out?

4.6.4 Explosives

¹H NMR spectroscopy can be used to study both pre- and post-blast explosives [40, 43]. The technique is not the choice for examining trace amounts, but with continuing improvements in instrument sensitivity there is scope to expand the use of this technique in the field of explosives. NMR spectroscopy is useful when there is a good quantity of material available for analysis as it can be used with prior separation of the components. Often post-blast samples can be extracted from debris using a solvent such as acetone. Samples can also be recovered for further analyses. An advantage of NMR spectroscopy is that analysis of relatively small amounts of impurities or additives can be carried out. The characterization of such components can provide a valuable means of identifying the source of the explosives.

4.7 Summary

Molecular spectroscopy is a valuable source of analytical techniques for forensic samples. Infrared spectroscopy is widely employed due to the versatility of sampling techniques. This vibrational spectroscopy technique can be used for paint, fibres, polymers, documents, explosives and drugs. Raman spectroscopy is often regarded as a complimentary technique to infrared spectroscopy and is utilized to examine drugs, paint, fibres, documents, explosives and GSR. UV–vis spectroscopy is a form of electronic spectroscopy that enables molecular processes occurring in the ultraviolet and visible regions of the spectrum to be utilized. This technique lends itself to the analysis of dyes and pigments in fibres, paint and questioned documents, as well as from drug and toxicological analysis. Fluorescence can also be examined using a spectroscopic method, and the fluorescence properties exhibited by body fluids, toxicological specimens and fibres can be used for qualitative or quantitative analysis. NMR spectroscopy is also used for structural determination, particularly organic structures, and its main application in forensic chemistry is in drug analysis, but it can also be useful for the analysis of explosive materials.

References

1. Chalmers, J. M., Edwards, H. G. M. and Hargreaves, M. D. (*Eds*), *Infrared and Raman Spectroscopy in Forensic Science*, John Wiley & Sons, Ltd, Chichester, 2012.
2. Ferrer, N., Forensic science, applications of IR spectroscopy, in *Encyclopedia of Spectroscopy and Spectrometry*, 2nd ed., Elsevier, Amsterdam, 2010, pp. 681–692.
3. Bartick, E. G., Applications of vibrational spectroscopy in criminal forensic analysis, in *Handbook of Vibrational Spectroscopy*, J. M. Chalmers and P. R. Griffiths (Eds), John Wiley & Sons, Ltd, Chichester, 2002, pp. 2994–3004.
4. Suzuki, E. M., Forensic applications of infrared spectroscopy, in *Forensic Science Handbook Vol. 3*, 2nd ed., R Saferstein (Ed.), Prentice Hall, Englewood Cliffs, NJ, 2010, pp. 75–251.
5. Debska, B. and Guzowska-Swide, E., Infrared spectral databases, in *Encyclopedia of Analytical Chemistry*, R. A. Meyers (Ed.), John Wiley & Sons, Ltd, Chichester, 2000, pp. 10928–10953.
6. Zadora, G., Chemometrics and statistical considerations in forensic science, in *Encyclopedia of Analytical Chemistry*, R. A. Meyers (Ed.), John Wiley & Sons, Ltd, Chichester, 2010.
7. Beveridge, A., Fung, T. and MacDougall, D., Use of infrared spectroscopy for the characterisation of paint fragments, in *Forensic Examination of Glass and Paint*, B. Caddy (Ed.), Taylor and Francis, London, 2001, pp. 183–242.
8. ASTM Standard E1610, *Standard Guide for Forensic Paint Analysis and Comparison*, American Society for Testing and Materials, West Conshohocken, PA, 2008.
9. Ryland, S., Infrared microscopy of forensic paint evidence, in *Practical Guide to Infrared Microscopy*, H. Humecki (Ed.), Marcel Dekker, New York, 1995, pp. 163–243.
10. Tungol, M. W., Bartick, E. G. and Montaser, A., Forensic examination of synthetic textile fibres by microscopic infrared spectrometry, in *Practical Guide to Infrared Microspectroscopy*, H. Humecki (Ed.), Marcel Dekker, New York, 1995, pp. 245–285.
11. Kirkbride, K. P. and Tungol, M. W., Infrared microspectroscopy of fibres, in *Forensic Examination of Fibres*, J. Robertson and M. C. Grieve (Eds), 1999, pp. 179–222.

12. ASTM Standard E2224, *Standard Guide for Forensic Analysis of Fibres by Infrared Spectroscopy*, American Society for Testing and Materials, West Conshohocken, PA, 2010.
13. Bartick, E. G., Tungol, M. W. and Reffner, J. A., *Anal. Chim. Acta* **288**, 35–42 (1994).
14. Merrill, R. A., Bartick, E. G. and Taylor, J. H., *Anal. Bioanal. Chem.* **376**, 1272–1278 (2003).
15. McNesby, K. L. and Pesce-Rodriquez, R. A., Applications of vibrational spectroscopy in the study of explosives, in *Handbook of Vibrational Spectroscopy*, J. M. Chalmers and P. R. Griffiths (Eds), John Wiley & Sons, Ltd, Chichester, 2006.
16. Monts, D. L., Singh, J. P. and Bourdreaux, G. M., Laser- and optical-based techniques for the detection of explosives, in *Encyclopedia of Analytical Chemistry*, R. A. Meyers (Ed.), John Wiley & Sons, Ltd, Chichester, 2010.
17. Zitrin, S., Analysis of explosives by infrared spectrometry and mass spectrometry, in *Forensic Investigation of Explosions*, A. Beveridge (Ed.), Taylor and Francis, London, 1998, pp. 267–314.
18. Bartick, E. G. and Buzzini, P., Raman spectroscopy in forensic science, in *Encyclopedia of Analytical Chemistry*, R. A. Meyer (Ed.), John Wiley & Sons, Ltd, Chichester, 2009.
19. Buzzini, P., Massonet, G. and Monard Sermier, F., *J. Raman Spect.* **37**, 922–931 (2006).
20. Kupstov, A. H., *J. Forensic Sci.* **39**, 305–318 (1994).
21. Buzzini, P. and Stoecklein, W., Forensic sciences / paints, varnishes and lacquers, in *Encyclopedia of Analytical Science*, P. Worsfield, A. Townshend and C. F. Poole (Eds), Elsevier, Amsterdam, 2005, pp. 453–464.
22. Lepot, L., de Wael, K., Gason, F. and Gilbert, B., *Science and Justice* **48**, 109–117 (2008).
23. Claybourn, M. and Ansell, M., *Science and Justice* **40**, 261–271(2000).
24. Moore, D. S., and Scharff, R. J., *Anal. Bioanal. Chem.* **393**, 1571–1578 (2009).
25. Ali, E. M. A., Edwards, H. G. M. and Scowen, I. J., *J. Raman Spect.* **40**, 2009–2014 (2009).
26. Worsfield, P. J., Spectrophotometry – overview, in *Encyclopedia of Analytical Sciences*, 2nd ed., P. Worsfold, A. Townshend and C. Poole (Eds), Elsevier, Amsterdam, 2005, pp. 318–321.
27. Eyring, M. B. Visible microscopical spectrophotometry in the forensic sciences, in *Forensic Science Handbook Vol. 1*, R. Saferstein (Ed.), 2nd ed., Prentice Hall, Upper Saddle River, NJ, 1993, pp. 321–387.
28. Adam, C., *Spectroscopy Europe* **2**, 13–16 (2009).
29. Zeiba-Palus, J., Microspectrophotometry in forensic science, in *Encyclopedia of Analytical Chemistry*, R. A. Meyer (Ed.), John Wiley & Sons, Ltd, Chichester, 2010.
30. Adolf, F. P. and Dunlop, J., Microspectrophotometry / colour measurement, in *Forensic Examination of Fibres*, J. Robertson and M. Grieve (Eds), CRC Press, Boca Raton, 1999, pp. 251–289.
31. Stoecklein, W., The role of colour and microscopic techniques for the characterisation of paint fragments, in *Forensic Examination of Glass and Paint*, B. Caddy (Ed.), CRC Press, New York, 2001.
32. Mills, T., Roberson, J. C., Matchett, C. C., Simon, M. J., Burns, M. D. and Ollis, R. J. (Eds), *Instrumental Data for Drug Analysis*, 3rd ed., CRC Press, Boca Raton, 2005.
33. Huddle, B. P. and Stephens, J. C., *J. Chem. Ed.* **80**, 441–443 (2003).
34. Menzel, E. R., Fluorescence in forensic science, in *Encyclopedia of Analytical Chemistry*, R. A. Meyers (Ed.), John Wiley & Sons, Ltd, Chichester, 2010.
35. Virkler, K. and Lednev, I. K., *Forensic Sci. Int.* **188**, 1–17 (2009).
36. Powers, L. S. and Lloyd, C. R., *US Patent* 7186990, 2004.
37. O'Reilley, J. E., *J. Chem. Educ.* **52**, 610–612 (1975).
38. Meloan, C. E., James, R. E., Saferstein, R. and Brettell, T., *Lab Manual for Criminalistics: An Introduction to Forensic Science*, Prentice Hall, Upper Saddle River, NJ, 2010.
39. Gaudette, B. D., The forensic aspects of textile fibre examination, in *Forensic Science Handbook Vol. 2*, R. Saferstein (Ed.), Prentice Hall, Englewood Cliffs, NJ, 1988.

40. Dawson, B., NMR spectroscopy applications – forensic, in *Encyclopedia of Analytical Science*, P. Worsfield, A. Townshend and C. F. Poole (Eds), Elsevier, Amsterdam, 2005, pp. 315–321.

41. Dawson, B., Nuclear magnetic resonance spectroscopy for the detection and quantification of abused drugs, in *Encyclopedia of Analytical Chemistry*, R. A. Meyers (Ed.), John Wiley & Sons, Inc., New York, 2006

42. Hays, P. A., *J. Forensic Sci.* **50** , 1342–1360 (2005).

43. Groombridge, C. J., NMR spectroscopy in forensic science, in *Annual Reports on NMR Spectroscopy Vol.* **32**, G. A. Well (Ed.), Academic Press, London, 1996, pp. 215–297.

Chapter 5

Elemental Analysis

Learning Objectives

- To understand atomic spectrometry techniques and how to obtain data from the techniques for forensic samples.
- To apply atomic spectrometry to the study of glass, gunshot residues and toxicological samples.
- To understand inductively coupled plasma–mass spectrometry (ICP–MS) and how to obtain data from the technique for forensic samples.
- To apply ICP–MS to the study of glass, gunshot residue and paint.
- To understand X-ray fluorescence spectroscopy and how to obtain data from the technique for forensic samples.
- To apply X-ray fluorescence spectroscopy to the study of glass, gunshot residue and paint.
- To understand particle induced X-ray emission spectroscopy and how to obtain data from the technique for forensic samples.
- To apply particle induced X-ray emission spectroscopy to the study of glass.

5.1 Introduction

The identification and quantification of elements in many types of evidence provide a valuable source of information in forensic chemistry. Elemental analysis can enable a sample to be linked to a crime scene or a perpetrator. Atomic spectroscopy techniques provide a sensitive means of elemental analysis. In atomic spectroscopy, a substance is decomposed with a flame, furnace or plasma

Forensic Analytical Techniques, First Edition. Barbara Stuart.
© 2013 John Wiley & Sons, Ltd. Published 2013 by John Wiley & Sons, Ltd.

and the concentrations of the species produced can be measured. The forms of atomic spectroscopy most commonly employed in forensic science are atomic spectrometry (including atomic absorption spectrometry and atomic emission spectrometry), inductively coupled plasma–mass spectrometry and X-ray fluorescence spectroscopy. An ion beam method, particle-induced X-ray emission spectroscopy, also enables elemental information to be obtained.

5.2 Atomic Spectrometry

The types of atomic spectrometry that are used for forensic applications are *atomic absorption spectrometry* (AAS) and *atomic emission spectrometry* (AES), with AAS instruments being more commonly available in laboratories [1, 2]. In AAS, atoms absorb a fraction of light from a source, while the remainder of the light from the source reaches the detector. In AES, the emission that results is from the loss of energy by atoms in a thermally excited state. As these techniques involve the measurement of absorbed or emitted radiation at a particular wavelength rather than recording spectra, the methods are known as atomic spectrometry. AAS and AES techniques provide a means of analysing a broad range of elements and can detect to low concentrations of the order of ppb to ppm.

5.2.1 Methods

In an atomic absorption spectrometer, a sample is atomized using either a flame or, in more modern instruments, a graphite furnace (graphite furnace atomic absorption spectrometry, or GFAAS). Figure 5.1 illustrates the layout of a GFAAS instrument. Flame AAS is relatively inexpensive compared to GFAAS, but larger quantities of sample (1–2 ml) are required for flame AAS, while volumes of only 5–10 μl or solids of 1–5 mg are required for GFAAS. Flame AAS allows concentrations of the order of ppm to be measured, while GFAAS enables a higher

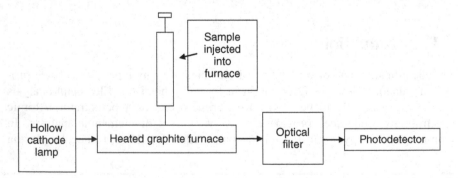

Figure 5.1 Layout of a GFAAS instrument.

sensitivity of the order of ppb. As elements absorb at different wavelengths, each element requires a specific wavelength produced by the source, known as a hollow cathode lamp. The intensity of radiation transmitted by the sample is measured during the experiment.

A traditional atomic emission spectrometer uses a flame to excite the atoms and is also commonly known as flame photometry. The dissolved sample is sprayed into a flame, and the intensity of emitted light is measured at selected wavelengths. The emission lines produced are used to identify elements. Due to the limitations and lack of sensitivity of traditional AES, the use of an inductively coupled plasma (ICP) to replace the flame has expanded. An ICP is produced by ionizing a flowing gas such as argon in a strong magnetic field. The temperature in the plasma can reach 10 000 K, enabling a broader range of elements to be analysed compared to a conventional flame. Concentrations of the order of ppb can be measured using this type of instrumentation compared to ppm obtained using a flame. An emerging AES technique is laser-induced breakdown spectroscopy (LIBS), which uses a high-energy pulsed laser as an excitation source. The laser is focussed on the sample surface to produce a plasma that atomizes the sample and results in the ablation of picogram to nanogram quantities of material. Fibre optics are used to collect the emission.

5.2.2 Interpretation

For AAS, the sample and a series of external standard solutions containing the elements of interest are analysed. The absorbance values are calculated using the intensity of the transmitted radiation for a sample and that for the solvent according to Equation 4.2. The absorbance values for the standard solutions are used to produce a calibration group of absorbance versus concentration. The graph will be linear as Beer's law is obeyed at low concentrations. Another method used to calculate a concentration in AAS is the standard addition approach. In this method, a known amount of a standard solution is added to the sample and the absorbance determined. The increase in the absorbance is due to the known added quantity, so the relationship between absorbance and concentration can be determined and, subsequently, the concentration of the sample solution can be calculated. In AES, the emission intensity is proportional to the amount of analyte present, so a calibration graph of emission intensity versus concentration may be produced. Thus, the emission intensity can be used to determine the analyte concentration from the calibration graph.

5.2.3 Glass

Both AAS and AES techniques are used for the analysis of forensic glass samples [1–5]. AAS is recognized as a fast and simple method for the analysis of single glass elements, with better sensitivity achieved with the use of GFAAS. ICP–AES also more readily enables multi-element analysis of glass to be carried out.

Clean glass samples are dissolved in hydrofluoric and hydrochloric acids for analysis. The elements of interest to be examined using atomic spectrometry include Sr, Ti, Fe. Ba, Mn, Ca, Al and Na.

SAQ 5.1

A 0.52 g glass fragment is analysed for Mg content using AAS. The fragment is dissolved in hydrofluoric acid–hydrochloric acid, and distilled water is added to make a volume of 1.00 l. An absorbance of 0.210 is measured for this solution. A standard addition approach is taken for the analysis of the glass, with 0.010 ml of a 1.00 g l^{-1} standard solution of Mg^{2+} added to 1.00 l of the unknown glass solution. The absorbance reading increases to 0.462 for the modified solution. Calculate the Mg concentration in the fragment.

5.2.4 Gunshot Residue

AAS is a straightforward technique for the elemental analysis of GSR primer residues [1, 6]. Samples are usually collected with cotton swabs and dissolved in nitric acid. The concentrations of Ba and Sb can be determined for hand swab samples using GFAAS. Conventional flame AAS has been shown to have inadequate sensitivity for the detection of these elements, but GFAAS has proved effective for detecting concentrations of the order of μg ml^{-1} for such elements.

In firearm cases where a firearm is not recovered or a bullet is too damaged to carry out testing based on physical properties, the use of elemental analysis can assist in linking evidence to a source. ICP–AES can be used to carry out bullet-casing analysis, where trace elements can be analysed to provide a means of identifying the bullet source [2]. The specimens can be dissolved in nitric acid for analysis. The elements Sb, Sn, Cd, As, Cu, Bi and Ag are measured and can provide a good discrimination of bullet specimens.

5.2.5 Toxicology

AAS and AES are both used for the analysis of metals in forensic toxicology [1, 7, 8]. GFAAS is a commonly applied technique as it provides suitable sensitivity. The tissue for analysis requires digestion in acid, with blood or urine specimens not necessarily requiring this procedure. Hair has been a widely studied material for the toxicological analysis using atomic spectrometry. Hair can be easier to work with compared to other matrices and contains suitable elemental concentrations.

SAQ 5.2

Outline a GFAAS method for the analysis of a blood sample believed to contain an elevated level of lead.

5.3 Inductively Coupled Plasma–Mass Spectrometry

Mass spectrometry (MS) is used to carry out qualitative and quantitative analysis of atoms, molecules and molecular fragments [9]. In MS, a gaseous sample is bombarded with high-energy electrons, which cause one or more electrons to be ejected on impact. A magnetic field is used to separate the ions by their masses. Molecular MS is examined in Chapter 6. In atomic MS techniques, the sample is atomized. The atoms are converted to a stream of ions, usually singly charged positive ions, which are separated on the basis of their mass-to-charge (m/z) ratio. The most useful form of atomic MS is inductively coupled plasma–mass spectrometry (ICP–MS), which has become one of the most important techniques for elemental analysis. ICP–MS can detect metals and some nonmetals at very low concentrations, by ionizing the sample with ICP and then analysing those ions.

5.3.1 Methods

In most ICP–MS instruments, a liquid sample is dispersed into a stream of gas (usually Ar or He) and injected into the core of an ICP. Figure 5.2 illustrates the layout of an ICP MS instrument. The sample is heated, vaporized and then ionized. The ions are passed through a core into a reduced pressure region and a number of ions pass through a second aperture into a high vacuum. Most instruments use a quadrupole mass analyser where the ions are detected. The transmitted ions are collected at a detector where the m/z values correspond to the element's natural isotope. The method is quantitative because the number of ions detected for each isotope will depend on the concentration of the element in the sample.

Generally samples are extracted and completely dissolved in acid, usually HNO_3, HCl and/or HF. Metals are the simplest to dissolve in acids, while glasses

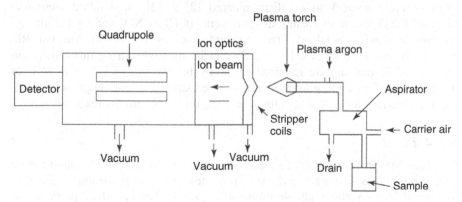

Figure 5.2 Layout of an ICP–MS instrument.

require more aggressive techniques involving higher temperatures and pressures. A typical ICP–MS instrument is able to detect concentrations of ng l^{-1}.

Although conventional ICP–MS instruments are designed to analyse liquids, laser ablation (LA) instruments are designed to analyse solids. LA–ICP–MS uses a laser to remove material from a small area of a sample, and the particles produced are removed from the sample in a flow of gas and injected into the plasma. Although the ablation process is destructive, the changes are on a microscopic scale. No sample preparation for LA–ICP–MS is required, but the sampling area should be flat.

5.3.2 Interpretation

An ICP–MS spectrum is a plot of the ion intensity versus the m/z ratio. As it is common for singly charged ions to be produced, the m/z ratio is commonly replaced by simply mass in the spectrum. The value of ICP–MS derives from the ability to carry out simultaneous multi-element quantitative analysis while using it. There are several calibration methods that may be employed. An external standard method, as described in Section 5.2, can be used and this is a straightforward approach when the system does not involve complex matrix effects. An internal standard method, which involves the addition of a reference material to which concentrations can be ratioed, overcomes the problems associated with mismatched matrices. The isotope dilution method involves the addition of standards of isotopes of the elements of interest. As the isotope has the same chemical properties as the analyte, it provides an accurate analytical approach. The standard addition approach, also described in Section 5.2, is useful for more difficult matrices.

5.3.3 Glass

ICP–MS methods can be used to provide elemental analysis that enables glass from different sources to be discriminated [2, 9–12]. A standard technique involves the digestion of clean glass fragments in HF, HNO_3 and HCl acids and an internal standard is added. Trace elements such as Mg, Al, Ti, Mn, Ga, Rb, Sr, Zr, Sb, Ba, La, Ce, Sm, Hf and Pb can be measured in ng quantities, and samples of several mg are required for analysis. The use of LA–ICP–MS for the analysis of forensic glass specimens is expanding as this approach avoids complete destruction of the specimen and the use of hazardous reagents.

5.3.4 Paint

LA–ICP–MS provides qualitative and semi-quantitative information about paint specimens and is favoured because it provides sensitive elemental analysis of individual layers without sample preparation [9, 13]. Time-resolved plots can be recorded by drilling a continuous crater through the layers, and changes in the elemental composition can be monitored. Quantitative analysis is more complex because obtaining matrix standards is difficult and paints are often heterogeneous in nature. Another approach is to determine elemental ratios where the relative intensities of particular elements are plotted.

5.3.5 Gunshot Residue

ICP–MS is a valuable tool when a bullet is damaged because lead isotope ratios, Sb/Pb ratios and trace element analyses can be carried out on fragments collected from a crime scene or from a victim's body to link to a source [2, 14]. Bullets can be produced from different lead alloys. The composition of a lead sample is based on a mixture four stable lead isotopes that are the products of natural decay series:

$$^{204}Pb \text{ (nonradiogenic)}$$

$$^{238}U \rightarrow {}^{206}Pb$$

$$^{235}U \rightarrow {}^{207}Pb$$

$$^{233}Th \rightarrow {}^{208}Pb$$

The isotopic abundances of lead isotopes vary and are dependent on the age of the ore used in to manufacture bullets. Thus, lead isotope ratios can be used to discriminate bullets based on regional variation. The ratios $^{208}Pb/^{206}Pb$ and $^{207}Pb/^{206}Pb$ are used. The Sb/Pb ratio also varies between bullets because antimony is a typical additive to modify the hardness of a bullet. As Sb is added in specific quantities for specific bullet types, the ratio can be used to discriminate samples. Analysis can be carried out to compare a sample with the composition of reference bullets from different manufacturers.

SAQ 5.3

In the case of a fatal shooting incident, particles from a bullet are recovered from the victim. Two weapons were discharged in the incident, and the weapon responsible for the fatality is damaged. ICP–MS analysis is used to carry out lead isotope analysis on the particles. The ratios determined for the retrieved particle and for bullets from the two weapons are listed in Table 5.1. It is possible to link the particle to a weapon used in the incident?

5.4 X-Ray Fluorescence Spectroscopy

X-ray fluorescence (XRF) spectroscopy is a widely used nondestructive technique for the measurement of the elemental composition of forensic materials [15–17].

Table 5.1 Lead isotope ratios for bullets (cf. SAQ 5.3)

Sample	$^{208}Pb/^{206}Pb$	$^{207}Pb/^{206}Pb$
bullet 1	2.002	0.816
bullet 2	2.046	0.835
particle	2.007	0.814

A sample is placed in a beam of high-energy photons produced by an X-ray tube. The radiation may be absorbed by an atom and transfer all of its energy to an innermost electron. If the primary X-ray has sufficient energy, electrons are ejected from the inner shells and create vacancies. Such vacancies produce an unstable atom and as the atom returns to a stable condition, the electrons from the outer shells transfer to the inner shells. During this process, a characteristic X-ray is emitted with an energy representing the difference between the two binding energies of the corresponding atomic shells. As each element contains a unique set of energy levels, each element produces X-rays with a unique set of energies.

5.4.1 Methods

The principal types of XRF spectrometers are wavelength dispersive and energy dispersive. While wavelength-dispersive XRF (WDXRF) measures wavelength, in energy-dispersive XRF (EDXRF), the energy of the fluorescent radiation is measured. WDXRF spectrometry requires more expensive instrumentation and needs larger sample sizes, making this technique less attractive for forensic applications. EDXRF does have the advantage that the instrumentation is simpler and less expensive. The simplification of the instrument means that portable instruments have been developed: while in standard instruments, an X-ray tube is used as a source, portable instruments can use radioisotope sources to allow for the rapid examination of samples. Micro-XRF spectrometers have also been developed, allowing for the examination of a very small area of a sample surface. These permit the study of the distribution of elements in the sample of interest. Such instruments may also be portable and allow the mapping of a surface. Figure 5.3 shows the layout of a typical XRF instrument.

5.4.2 Interpretation

Generally the innermost K and L shells are involved in XRF (Figure 5.4). For the production of K lines in XRF, an electron from the L or M shell falls to fill the vacancy, emitting a characteristic X-ray and, in turn, producing a vacancy in the L or M shells. In the production of L lines, when a vacancy is created in the L shell by either the primary X-ray or a previous event, an electron from the M or N shell moves to occupy the vacancy. In this process, it emits a characteristic X-ray and, in turn, produces a vacancy in the M or N shell. The characteristic X-rays

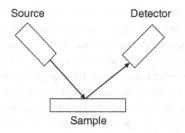

Figure 5.3 Layout of an XRF instrument.

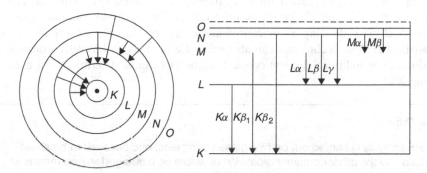

Figure 5.4 Transitions observed in XRF. Reproduced with permission from L. Moens et al., in *Modern Analytical Methods in Art and Archaeology*, eds. E. Ciliberto and G. Spoto, Wiley, New York (2000), pp 55-79.

produced in XRF are labelled K, L, M or N, denoting the shells from which they originate. Where X-rays originate from the transitions of electrons from higher shells, an α, β or γ designation is used to label the X-rays. A further designation is made as $\alpha_1, \alpha_2, \beta_1, \beta_2$ and so on to represent the transition of electrons from the orbits of higher and lower binding energy electrons within the shells into the same lower shell.

A spectrum is produced in XRF spectroscopy usually showing counts as a function of binding energy. The spectra are compared with standards for identification purposes. XRF spectroscopy works well for the analysis of elements with higher atomic numbers. Qualitative and quantitative analysis can be carried out to determine amounts of the order of ppm, depending on the instrument and sample type.

5.4.3 Glass

XRF is used to analyse glass evidence because it is a straightforward nondestructive technique capable of analysing multiple elements [2–5, 16]. EDXRF and micro-XRF techniques are often used as forensic glass specimens can be quite small. Specimens are usually cleaned with nitric acid and demineralized water added prior to analysis. A major issue for the analysis of glass is specimen geometry: a flat sample for X-ray contact is required for reproducible results. However, the result for specimens can be compared with those for standards of similar geometry or by embedding in resin and polishing to produce a flat surface. XRF can quantify a broad range of elements present in glasses and as XRF is sensitive to higher atomic numbers, it is useful for quantifying elements such as Mn, Sr and Zr, which can be useful for discrimination of glasses. Quantitative analysis of samples is best achieved by measuring the ratios of elemental concentrations, rather than the measurement of absolute concentrations. For example, different Ca/Fe ratios are observed for sheet and container glasses, with sheet glass showing a low ratio while container glass shows a higher ratio. Statistical analysis techniques can also be applied to handle the concentrations of major, minor and trace element concentrations that can be determined for glasses using XRF.

SAQ 5.4

XRF analysis is carried out on two glass specimens, one believed to be sheet glass and the other container glass. What would be a straightforward means of classifying the specimens using just XRF data?

5.4.4 Gunshot Residue

XRF provides a means of determining the elemental composition of GSR, with elements including K, Fe, Cu, Ba and Pb commonly used to discriminate particles [16, 18]. Due to the small size of GSR particles, micro-XRF provides an effective approach. Figure 5.5 shows an example of a micro-XRF spectrum obtained for a primer residue on cloth. The technique can also be used to produce an elemental distribution map to obtain a picture of the distribution of GSR particles on a surface, which is useful for determining the spread of residue.

5.4.5 Paint

XRF provides elemental information about pigments and fillers and so enables paint specimens, such as those collected from vehicles, to be characterized and potentially linked to a crime scene [16, 19, 20]. If a traditional XRF approach is used, information is obtained about the composition of more than one paint layer. If information about individual layers is required, micro-XRF can be used to carry out elemental analysis on separate layers, thus providing additional discriminating information.

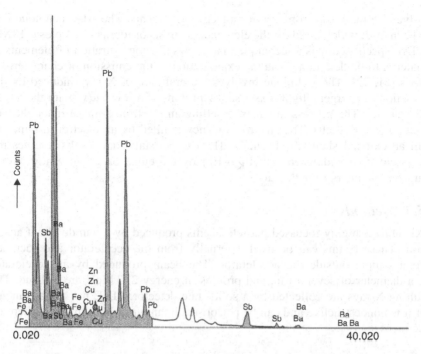

Figure 5.5 Micro-XRF spectrum of a GSR primer residue. Reproduced from J. Flynn et al., *Forensic Sci. Int.* **97**, 1, 21-36 with permission from Elsevier Ireland Ltd (1998).

SAQ 5.5

XRF analysis is carried out on a yellow-coloured paint smear collected from the surface of a vehicle believed to be involved in a collision. The major elements detected in the paint smear are Ti, Pb and Cr. What does this analysis reveal about the paint smear composition?

5.5 Particle-Induced X-Ray Emission Spectroscopy

Ion beam analysis (IBA) involves the irradiation of a sample by an ion beam and the detection of emitted radiation or particles that provide structural information. There is a number of types of IBA, including particle-induced X-ray emission (PIXE) spectroscopy, Rutherford back-scattering spectrometry (RBS), nuclear reaction analysis (NRA) and secondary ion mass spectrometry (SIMS). The use of SIMS to forensic science will be covered in Chapter 6 as this technique involves the mass spectrometry of molecular fragments. The IBA technique that has been most widely used for the elemental analysis of forensic samples is PIXE.

PIXE spectroscopy is a technique that allows the concentration of elements in a material to be determined via the examination of the emission of characteristic X-rays [21, 22]. The technique involves the emission of X-rays induced by the interaction of energetic light ions, usually protons of a few MeV, with the atoms in a material. The process involves expelling an electron from an inner shell of the atom (the K shell). The electron vacancy is filled by an electron originating from an external shell (the L shell). The excited state that results releases the excess electron-binding energy $(E_K - E_L)$ by the emission of an X-ray with a characteristic energy for the atom.

5.5.1 Methods

PIXE utilizes highly focussed particle beams produced by a van de Graaf accelerator. These beams can be used externally from the accelerator and focussed onto a sample outside the accelerator. The beam produced by an accelerator has a diameter of several μm, and protons of energy 2–3 MeV are utilized. The resulting X-rays are collected by a solid-state detector. PIXE has the advantage that it is nondestructive and can be performed in an air or helium atmosphere on specimens.

5.5.2 Interpretation

In PIXE, elements with an atomic number greater than 11 can be detected simultaneously by their K or L lines. The detection limit is in the μg g^{-1} range, making PIXE suitable for trace element analysis. The data are typically represented by a spectrum showing responses as a function of the X-ray energies.

DQ 5.1

Can PIXE be classed as a destructive or nondestructive technique?

Answer

Although PIXE can slightly change a sample surface due to the interaction of the beam, the technique is widely perceived as nondestructive as such changes are very minor. As no modification to a specimen is required prior to PIXE analysis, the specimen can be analysed further using other techniques.

5.5.3 Glass

Although PIXE is a less accessible elemental technique than, say, XRF, it can provide valuable quantitative information about the heavier elements that can be present in glass [21]. Figure 5.6 provides an example of a PIXE spectrum of a

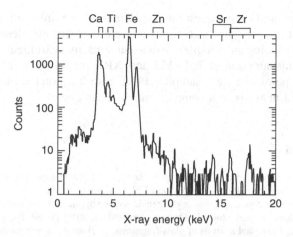

Figure 5.6 PIXE spectrum of a glass fragment. Reproduced with permission from P.A. Young, C.C. Hall, P.J. Mears, D.J. Padilla, R. Sampson and G.F. Peasloe, 'Comparion of glass fragments using particle induced x-ray emission (PIXE) spectrometry', *J. Forensic Sci*. **56**, 366-371 with permission from John Wiley & Sons, (2011).

glass fragment where Ca, Ti, Fe, Zn, Sr and Zr are detected. For quantitative analysis the data can be normalized by a comparison with a known standard, such as a National Institute of Standards and Technology (NIST) glass standard.

With the advent of microbeam PIXE analysis, there is scope for this technique to be applied to trace evidence, such as GSR and soil, where discrimination is not achieved via other elemental techniques.

5.6 Neutron Activation Analysis

A brief description of another method of elemental analysis, neutron activation analysis (NAA), is made here. NAA is a technique that involves bombarding a nonradioactive sample with neutrons. A small fraction of the atoms are converted to radioisotopes, and the characteristic decay patterns are recorded to identify the elements present. Until the advent of ICP–MS and PIXE, NAA was a standard analytical method for elemental analysis at the ppm level. Although NAA offers high sensitivity and minimal sample preparation, this approach requires access to a nuclear reactor, rendering it unsuitable for routine analysis. NAA has been used for the examination of GSR, as it is suitable for the determination of barium and antimony [23, 24]. However, as NAA is unsuitable for lead determination, it is less useful than other more accessible elemental techniques.

5.7 Summary

Atomic spectroscopy provides a number of sensitive techniques that are able to analyse multiple elements in forensic samples. The techniques of AAS, AES, ICP–MS and XRF all provide a means of characterizing the content of a variety of material types and enable questioned specimens to be linked to a source. AAS and AES are straightforward techniques and are used for the elemental analysis of GSR and toxicological samples. Advancements in instrumentation have led to the increasing adoption of ICP–MS and XRF for the elemental analysis of forensic glass, paint and GSR samples. PIXE also has potential as an approach to trace element analysis as a complementary technique.

References

1. Grant, D. M. and Peters, C. A., Atomic spectroscopy for forensic applications, in *Encyclopedia of Analytical Chemistry*, John Wiley & Sons, Ltd, Chichester, 2006.
2. Duckworth, D. C., Atomic spectroscopy: forensic science applications, in *Encyclopedia of Spectroscopy and Spectrometry*, 2nd ed., Elsevier, Amsterdam, 2010, pp. 84–90.
3. Almirall, J. R., Elemental analysis of glass fragments, in *Trace Evidence Analysis and Interpretation: Glass and Paint*, B. Caddy (Ed.), Taylor and Francis, London, 2001, pp. 65–83.
4. Buscaglia, J. A., *Anal. Chim. Acta* **288**, 17–24 (1994).
5. Koons, R. D., Buscaglia, J., Bottrell, M. and Miller, E. T., Forensic glass comparisons, in *Forensic Science Handbook Vol*. 1, R. Saferstein (Ed.), Prentice Hall, Upper Saddle River, NJ, 2002, pp. 161–214.
6. Dalby, O., Butler, D. and Birkett, J. W., *J. Forensic Sci.* **55**, 924–943 (2010).
7. Jeckells, S. and Negrusz, A. (Eds), *Clarke's Analytical Forensic Toxicology*, Pharmaceutical Press, London, 2008.
8. Levine, B. (Ed.), *Principles of Forensic Toxicology*, AACC Press, Washington, DC, 2006.
9. Trejos, T. and Almirall, J. R., Laser ablation inductively coupled plasma mass spectrometry in forensic science, in *Encyclopedia of Analytical Chemistry*, R. A. Meyers (Ed.), John Wiley & Sons, Ltd, Chichester, 2010.
10. ASTM Standard E2330, *Standard Test Method for Determination of Trace Elements in Glass Samples Using Inductively Coupled Plasma Mass Spectrometry (ICP–MS)*, American Society for Testing and Materials, West Conshohocken, PA, 2004.
11. Trejos, T., Montero, S. and Almirall, J. R., *Anal. Bioanal. Chem.* **376**, 1255–1264 (2003).
12. Duckworth, D. C., Morton, S. J., Bayne, C. K., Montero, S., Koons, R. D. and Almirall, J. R., *J. Anal. Atom. Spectrom.* **17**, 662–668 (2002).
13. Hobbs, A. L. and Almirall, J. R., *Anal. Bioanal. Chem.* **376**, 1265–1271 (2003).
14. Ulrich, A., Moor, C., Vonmont, H., Jordi, H. R. and Lory, M., *Anal. Bioanal. Chem.* **378**, 1059–1068 (2004).
15. Fischer, R. and Hellmeiss, G., Principles and forensic applications of X-ray diffraction and X-ray fluorescence, *Advances in Forensic Science Vol*. 2, Medical Publishers, Chicago, 1989, pp. 129–158.
16. Roux, C. and Lennard, C., X-ray fluorescence in forensic science, in *Encyclopedia of Analytical Chemistry*, R. A. Meyers (Ed.), John Wiley & Sons, Ltd, Chichester, 2006.
17. Vittiglio, G., Bichmeier, S., Klinger, P., Heckel, J., Fuzhong, W., Vincze, L., Janssens, K., Engstrom, P., Rindlby, A., Dietrich, K., Jembrih-Simburger, D., Schreiner, M., Denis, D., Lakdar A. and Maotte, A., *Nucl. Inst. Methods Phys. Res. B* **213**, 693–698 (2004).
18. Flynn, J., Stoilovic, M., Lennard, C., Prior, I. and Kobus, H., *Forensic Sci. Int.* **97**, 21–36 (1998).
19. ASTM Standard E1610, *Standard Guide for Forensic Paint Analysis and Comparison*, American Society for Testing and Materials, West Conshohocken, PA, 2002.
20. Fitzgerald, S., *Spectroscopy Europe* **21**, 16–18 (2009).
21. DeYoung, P. A., Hall, C. C., Mears, P. J., Padilla, D. J., Sampson, R. and Peasloe, G. F., *J. Forensic Sci.* **56**, 366–371 (2011).

22. Jisonna, L. J., DeYoung, P. A., Ferens, J., Hall, C., Lunerberg, J. M., Mears, P., Padilla, D., Peasloe, G. F. and Sampson, R., *Nucl. Instr. Methods Phys. Res. B* **269**, 1067–1070 (2011).
23. Lewis, S. W., Agg, K. M., Gutowski, S. J. and Ross, P., Forensic sciences / gunshot residues, in *Encyclopedia of Analytical Science*, P. Worsfield, A. Townshend and C. F. Poole (Eds), Elsevier, Amsterdam, 2005, pp. 430–436.
24. Dalby, O., Butlerm D. and Birkett, J. W., *J. Forensic Sci.* **55**, 924–943 (2010).

Chapter 6

Mass Spectrometry

Learning Objectives

- To understand the origins of mass spectrometry techniques and how to obtain data from such techniques for forensic samples.
- To apply mass spectrometry to the study of drugs and explosives.
- To understand the origins of ion mobility spectrometry and how to obtain data from the technique for forensic samples.
- To apply ion mobility spectrometry to the study of explosives and drugs.

6.1 Introduction

Mass spectrometry (MS) is a widely used approach for the identification of unknown compounds. MS enables complex molecules to be characterized, even when only very small quantities of sample are available. MS techniques can be classified as atomic or molecular. Atomic MS was described in Chapter 5, and the molecular MS techniques commonly employed in forensic science are covered in this chapter. A related ion spectrometry technique, ion mobility spectrometry, is also described here.

6.2 Molecular Mass Spectrometry

MS is used to determine the masses of molecules or molecular fragments [1–3]. In MS, a gaseous sample is bombarded with high-energy electrons that cause one

Forensic Analytical Techniques, First Edition. Barbara Stuart.
© 2013 John Wiley & Sons, Ltd. Published 2013 by John Wiley & Sons, Ltd.

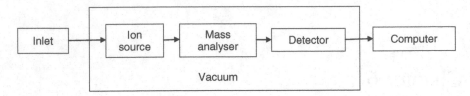

Figure 6.1 Layout of a mass spectrometer.

or more electrons to be ejected on impact. A magnetic field is used to separate the ions based on the masses. A number of MS techniques have been developed in order to deal with a range of sample types.

6.2.1 Methods

The fundamental components of a mass spectrometer are shown in Figure 6.1. The sample is introduced either directly or via a chromatographic inlet. Direct insertion enables solid specimens to be analysed and will be discussed in this section. The combination of chromatographic techniques with MS will be covered in Chapter 7. Once the sample is introduced into the instrument, an ionization source converts the sample to a gas. The molecules are then converted to ions which are then passed to a mass analyser, that separates the ions according to their mass-to-charge (m/z) ratio. The ions are focused onto a detector, usually an electron multiplier, by varying the field strength.

A number of ionization methods are commonly employed to deal with forensic specimens. *Electron ionization* (EI) is a common form of ionization used in MS and is carried out using electrons with an energy of 70 e V. The molecule loses an electron in the process, and a positively charged molecular ion (M^+) is formed. *Chemical ionization* (CI) involves bombarding the sample with positively charged atoms or molecules instead of electrons. CI is a milder technique that enables molecules unable to cope with EI conditions to be investigated. *Electrospray ionization* (ESI) is also a milder technique that passes a sample solution through a needle at an electrical potential of about 4 kV, allowing the sample to be directly taken from solution to ions in a gas. *Atmospheric pressure chemical ionization* APCI uses a corona discharge into an electrospray mist to produce a spark that ionizes the molecules. APCI is also a soft ionization technique, but is more energetic than ESI.

Two more recently developed ambient ionization sources enable solid specimen surfaces to be sampled without prior separation or preparation: *desorption electrospray ionization* (DESI) and *direct analysis in real time* (DART). In DESI, charged droplets are sprayed on the sample surface to create a thin film of liquid in which the analytes dissolve; this film is then drawn into the spectrometer. In DART, an electrical potential is applied to produce plasma that contains excited atoms and ions, and the plasma is then heated and applied to the sample surface. Analyte ions produced are desorbed into the gas phase and carried into the spectrometer.

Another surface-sensitive technique that may be employed is *secondary ion mass spectrometry* (SIMS). This is an ion beam technique (introduced in Chapter 5) and involves bombardment of the sample surface with an ion beam within an ultra-high vacuum $(1.33 \times 10^{-8}\,Pa)$ environment. Inert gas ions or metal ions such as Ga^+ or In^+ are used. The impact of the ion initiates a phenomenon known as a collision cascade in which the atoms in an approximately $10^3\,nm^3$ volume around the ion are in rapid motion. Some of the energy returns to the surface to break bonds and produce atomic and molecular species. A proportion of these are ionized during the emission process and are known as secondary ions. Secondary ions for analysis are generated from within 20 Å of the surface as these have sufficient energy to escape. SIMS can be used in static mode or dynamically where the ion beam is rastered across the sample surface and the secondary ion intensity as a function of position is recorded.

There are various mass analysers used in mass spectrometers in forensic laboratories. A *quadruple mass spectrometer* is one of the most common and involves the use of four rods that produce a voltage along the path of the ions. When direct-current (DC) and radiofrequency (RF) voltages are applied to the rods, the ions oscillate. For each *m/z* ratio, a stable oscillation exists that allows the ions to travel the entire length of the rods without being lost, enabling only ions of a specific *m/z* ratio to be collected. An *ion trap spectrometer* uses closed cavity electrodes that enable the molecular reactions to occur with a confined space and, therefore, increase their probability of occurrence. Ion trap instruments can be used to study unstable molecules. A *time-of-flight* (TOF) instrument is based on the principle that lighter ions are accelerated faster than heavier ions and so will have short TOFs over a specified distance. In a TOF spectrometer, the ions pass through a region with no field applied and the time to arrive at the detector is measured. Mass spectrometers can be combined with other apparatus to further their analytical capabilities. *Tandem MS* (or MS–MS) links at least two stages of mass analysis. A first spectrometer is used to isolate the component ions in a mixture and the ions are then introduced one at a time into a second spectrometer, where they are fragmented to produce a series of spectra.

As it is not always possible to directly analyse all forensic specimen types, consideration should be given to any necessary sample preparation. Analytes may need to be separated from a matrix (e.g. a drug molecule in urine) in order to avoid interference from the matrix molecules. Simple solvent extractions may be used where the analytes are dissolved in an appropriate solvent. Solid-phase extraction

(SPE), which involves the separation of analytes and a matrix using an appropriate stationary phase in a cartridge, can also be employed. Pyrolysis, where a sample is heated under controlled conditions to produce a gaseous sample, can also be used. The use of pyrolysis is covered in Chapter 8.

6.2.2 Interpretation

A mass spectrum produced in MS will show the relative abundance of the ions formed versus the m/z ratio. The base peak is the most abundant ion observed and is fixed at 100% relative abundance. Positive ions are usually studied, but negative ions can also be examined. When EI ionization is used, the most intense peak observed in the spectrum is due to the molecular ion ($M^+ \cdot$) – the molecule under study has lost an electron to form a cationic radical species. Any fragment ions produced will show peaks at lower mass values.

For the identification of an analyte with well-known fragmentation patterns, it is possible to use selected ion monitoring (SIM). Several selected ions are chosen as target compounds, and the ratios of these compounds are compared to a standard. Alternatively, in full-scan mode, a complete mass spectrum can be compared to a mass spectral library. For quantitative analysis, an internal standard method can be used.

6.2.3 Drugs

The identification of drugs is regularly carried out using MS techniques [1, 4–7]. The use of MS in combination with gas or liquid chromatography will be described in detail in Chapter 7, but it is also feasible to use direct-probe EI and CI MS for drug analysis. The advantage of this approach is that a specimen may be quickly analysed without the need for time-consuming sample preparation. The fragmentation patterns resulting from CI MS are simpler than those resulting from EI analysis. CI mass spectra of drug mixtures will usually contain a strong MH^+ and major fragmentation ions that can be assigned to various drug components. The strong MH^+ peak means that the molecular mass of a drug can be used for identification purposes. Table 6.1 lists the major ions detected in the mass spectra of common drug components. The spectra can be compared to a standard drug spectrum or a library database to provide a potential identification of the drugs present [8]. The assignment should be regarded as tentative because compounds of similar molecular masses can yield spectra that may be difficult to distinguish.

SAQ 6.1

One of the metabolites of heroin is 6-monoacetylmorphine, the structure of which is shown in Figure 6.2. In the CI mass spectrum of an illicit heroin sample, where would the molecular ion peak for this metabolite appear in the spectrum?

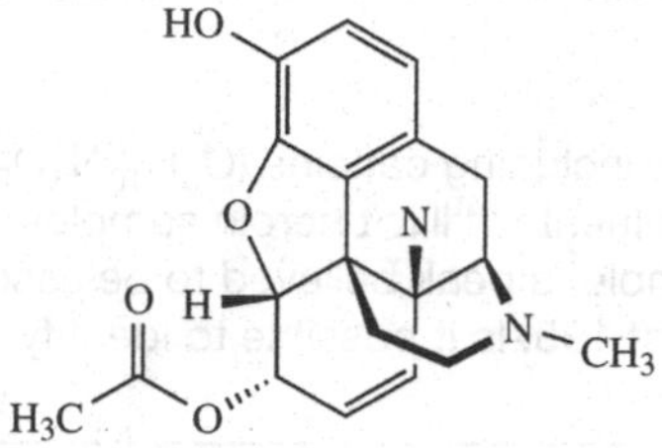

Figure 6.2 Structure of 6-monoacetylmorphine.

Table 6.1 Major ions in mass spectra for common drugs and precursors

Compound	Molecular mass	Base ion	Other major ions
amobarbital	226	156	141, 157, 142, 197, 98, 183
amphetamine	135	44	91, 92, 65, 120, 134, 77
cocaine	303	82	182, 83, 77, 94, 105, 303
codeine	299	299	162, 42, 115, 229, 214, 124
diazepam	284	256	283, 284, 221, 77, 165, 151
ephedrine	165	58	77, 51, 79, 105, 131, 56
flunitrazepam	313	285	312, 266, 238, 183, 248, 109
heroin	369	327	268, 369, 204, 310, 215, 43
ketamine	237	180	182, 209, 152, 138, 102, 115
lysergic acid diethylamide (LSD)	268	268	224, 207, 192, 154, 180, 167
mescaline	211	182	167, 181, 211, 151, 148, 136
methadone	309	72	91, 165, 115, 223, 178, 105
methamphetamine	149	58	91, 65, 56, 134, 115, 119
morphine	285	285	162, 215, 124, 115, 174, 268
pentobarbital	226	156	141, 157, 98, 197, 69, 112
phencyclidine	243	200	91, 242, 243, 186, 166, 84
phenobarbital	232	204	117, 232, 161., 146, 174, 103
pseudoephedrine	165	58	77, 79, 105, 42, 117, 131
psilocybin	284	58	204, 159, 146, 130, 183, 117
psilocin	204	58	204, 42, 77, 117, 146, 159
temazepam	300	271	273, 255, 300, 77, 165, 193
Δ^9-tetrahydrocannabinol (THC)	314	299	231, 314, 271, 243, 258, 91

SAQ 6.2

Various drug compounds, including caffeine ($C_8H_{10}N_4O_2$) and paracetamol ($C_8H_9NO_2$), are used as diluents in illicit heroin samples. In a CI mass spectrum collected for a heroin sample, a peak believed to be associated with a diluent is observed at an *m/z* ratio of 195. Is it possible to identify the compound used in this case?

SIMS shows potential as an approach to the identification of drugs present in trace quantities in fingerprints [9]. Lifted fingerprints may be examined using TOF-SIMS to produce images of the fingerprint surface. A corresponding micro-analysis can be used to demonstrate the presence of a particular drug that can be linked to a crime scene. The method can also be extended to the detection of trace quantities of other exogenous contaminants found in fingerprints, such as gunshot residue (GSR) and explosives, and can be used to examine materials on other difficult surfaces.

6.2.4 Explosives

A range of MS techniques can be employed to identify and characterize explosive materials [1, 10–13]. The detection limit of explosives using MS is of the order of picograms. Current commonly used ionization methods for the analysis of explosives include ESI and APCI. The methods used for explosives often involve the formation of cluster or adduct ions for identification purposes.

Figure 6.3 provides an example of a mass spectrum of an explosive, illustrating the negative-ion APCI mass spectrum of nitroglycerine. The ambient ionization DESI and DART techniques are also used in explosives analysis and are leading to the development of portable mass spectrometers that will enable MS to be used for detection in the field.

SAQ 6.3

An ESI produces a mass spectrum for an unknown explosive that has a major ion appearing at *m/z* = 226. If this peak is an [M − H]⁻ ion, is it possible to identify the explosives as one of those shown in Figure 1.5?

6.3 Isotope Ratio Mass Spectrometry

Stable isotopes are isotopes that do not decay via radioactive processes over time, and most elements consist of more than one stable isotope. The stable isotope content of a substance can provide valuable information about the origins of a particular sample, such as identifying the source of a material. Variations in isotopic composition can be accurately measured using isotope ratio mass spectrometry (IRMS) [14–18].

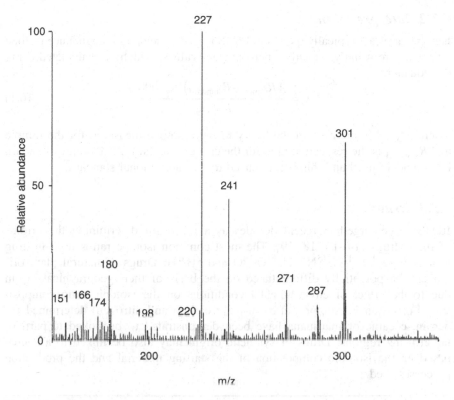

Figure 6.3 Negative-ion APCI mass spectrum of nitroglycerine. Reproduced from C.S. Evans et al., *Rapid Commun. Mass Spect.* **16**, 1883–1891 with permission from John Wiley & Sons (2002).

6.3.1 Methods

Isotope ratio mass spectrometers are designed to measure low isotopic concentrations and use fixed multiple detectors for different isotopomers. Samples are usually combusted to simple gaseous molecules such as CO_2 and H_2O prior to analysis. Typically the gas is introduced into an ionizing chamber and accelerated through a magnetic field. The ions are focused onto dedicated Faraday cups positioned for the masses of interest, and the ion currents recorded. For instance, if carbon isotopes are to be measured, three Faraday cups would be used to measure m/z values of 44, 45 and 46 as the primary species are $^{12}C^{16}O^{16}O$ (44), $^{13}C^{16}O^{16}O$ (45) and $^{12}C^{18}O^{16}O$ (46). The elements and their isotopes of interest in forensic science are hydrogen (^{1}H, ^{2}H), carbon (^{12}C, ^{13}C), nitrogen (^{14}N, ^{15}N), oxygen (^{16}O, ^{17}O, ^{18}O) and sulfur (^{32}S, ^{33}S, ^{34}S, ^{36}S). IRMS can be combined with a gas chromatograph (described in Chapter 7) to separate sample components and enhance the information provided.

6.3.2 Interpretation

Isotope ratios are typically quoted in IRMS results. The natural abundance isotope ratio data are usually quoted as percentage δ values, which are calculated using the equation:

$$\delta\% = \frac{(R_{\text{sample}} - R_{\text{standard}}) \times 1000}{R_{\text{standard}}} \tag{6.1}$$

where R_{sample} is the ratio of the heavy to light isotope measured for the sample and R_{standard} is the respective ratio for the chosen standard. A laboratory standard is commonly used and this is referenced to an international standard.

6.3.3 Drugs

IRMS has emerged in recent decades as a means of determining the origins of illicit drugs [14–16, 18, 19]. The most common isotope ratios used in drug profiling are $^{13}C/^{12}C$, $^{15}N/^{14}N$, $^{18}O/^{16}O$ and $^{2}H/^{1}H$. Drugs of natural plant origin can be potentially differentiated on the basis of their geographical origin due to the effect of environmental conditions on the isotopic ratio composition. For instance, isotopic ratios (usually carbon and nitrogen) determined for heroin, cocaine or marijuana have been demonstrated to be linked to particular growing regions. Drugs of synthetic origin may also be linked to a source based on the isotopic composition of the starting material and the production processes used.

SAQ 6.4

A cocaine sample is analysed using IRMS, and a $^{13}C/^{12}C$ ratio of 0.01085 is measured. The standard ratio is 0.01124. Determine the δ % value for this cocaine sample.

6.3.4 Explosives

IRMS can be used to discriminate a range of explosive materials [14, 16, 19]. The technique is useful for linking an explosive to its starting material, and potentially its source, as the isotopic composition can be connected to the reagents and the manufacturing process. Carbon and nitrogen isotopes are valuable as they are very commonly found in a wide range of explosives.

DQ 6.1

What sort of isotopes would be recommended choice(s) for an IRMS analysis of a sample of ammonium nitrate?

Answer

Given that the structure of ammonium nitrate is NH_4NO_3, the nitrogen, oxygen and hydrogen isotopes should be investigated.

6.4 Ion Mobility Spectrometry

Ion mobility spectrometry (IMS) is a technique used to detect trace quantities of gases and vapours [10, 20–23]. In IMS, a gaseous sample is ionized and different ions present are separated under the influence of an electric field as they travel at different velocities (mobilities) through a carrier gas.

6.4.1 Methods

The layout of a conventional IMS instrument is shown in Figure 6.4. A gaseous sample is drawn into an ionization chamber using a pump. There are a variety of possible ionization techniques, but the use of radionuclides (e.g. ^{63}Ni or ^{241}Am) is still common in instruments due to their long life and low maintenance. Positive or negative ions may be produced. The ions are then passed into a drift tube containing a carrier gas and are subjected to an electric field under atmospheric pressure. The flight times of the ions are detected. For forensic applications, the use of handheld and compact IMS devices is very popular. More recent instruments have involved the combination of IMS with mass spectrometry and/or chromatographic techniques.

6.4.2 Interpretation

In IMS, the time taken for an ion to traverse the distance to the detector is proportional to the mass of the molecule. When an ion passes through a gas under

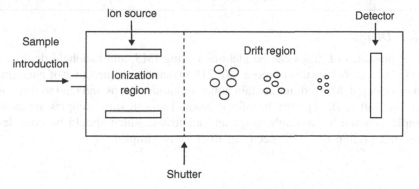

Figure 6.4 Layout of an IMS instrument.

the influence of an electric field, the drift velocity (v_d) is proportional to the field strength of the field (E):

$$v_d = K\,E \tag{6.2}$$

where K is the ion mobility. Ion mobilities are commonly reported as reduced mobilities, which takes into account factors such as the charge, mass and shape of the molecule. The reduced mobility can be calculated using the drift time, the length of the drift region and the electric field strength and corrected to a standard pressure and temperature. However, due to the lack of suitable reference materials, the reduced mobility values quoted for the same compound can vary. Mobility spectra are a plot of the signal intensity versus the drift time. The mobility spectra are typically compared to databases of samples of interest.

6.4.3 Explosives

IMS is a very common technique for the detection of explosives [10, 20, 21, 23, 24]. The technique is used in the field, such as at airports for the detection of hidden explosives or for the identification of post-blast debris. Many explosives contain nitro functional groups, and negative ions are usually formed via various mechanisms (e.g. charge transfer or proton transfer). A common reactant ion in air is O_2^-. An improvement in sensitivity can be obtained by the use of a dopant (e.g., methylene chloride to produce Cl^- ions): the more ions that are formed, the greater the likelihood of a positive identification. Figure 6.5 illustrates IMS spectra for various explosives.

SAQ 6.5

A suspected explosive specimen is analysed using IMS under the same conditions as those used to obtain the spectra shown in Figure 6.5. The resulting spectrum shows notable peaks corresponding to drift times in a 7–8 ms range. Is it possible to identify the explosive type present in this specimen?

6.4.4 Drugs

Trace quantities of drugs can be identified using IMS, and handheld devices are particularly useful for this purpose [20, 21]. Quantities of the order of nanograms can be detected. Many drugs contain amine or amide groups and tend to form positive ions, often via a proton transfer reaction. Forensic drug samples are usually complex, containing multiple drugs and additives, which should be considered when interpreting the IMS spectra of illicit drug samples.

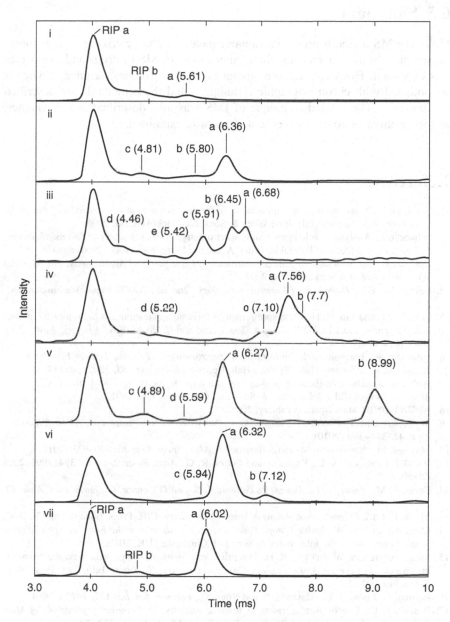

Figure 6.5 IMS spectra of explosives (i) 2,4-DNT; (ii) 2,6-DNT; (iii) NG; (iv) PETN; (v) RDX; (vi) tetryl; (vii) TNT (reactant ion peak). With kind permission from Springer+Business Media J.S. Babis et al., *Anal. Bioanal. Chem.* **395**, 411–419 with permission from Springer (2009).

6.5 Summary

Molecular MS methods provide a valuable means of characterizing various forensic samples. In this chapter, the direct application of MS to drugs and explosives was examined. However, the real strength of MS in forensic science is when it is combined with chromatographic techniques, and these methods are described in Chapter 7. The related technique of IMS was also described in this chapter, and its application to explosives and drugs was examined.

References

1. Smith, D. L., Mass spectrometry applications in forensic science, in *Encyclopedia of Analytical Chemistry*, R.A. Meyers (Ed), John Wiley & Sons, Inc., Hoboken, NJ, 2006.
2. Schubeth, J., Analytical techniques / mass spectrometry, in *Encyclopedia of Forensic Sciences*, J. Siegel, G. Knupfer and P. Saukko (Eds), Academic Press, New York, 2000, pp. 155–161.
3. Hoffman E. and Stroobant, V., *Mass Spectrometry: Principles and Applications*, 3rd ed., John Wiley & Sons Ltd, Chichester, UK, 2007.
4. Levine, B. (Ed.), *Principles of Forensic Toxicology*, 2nd ed., AACC Press, Washington, DC, 2006.
5. Lewis, R. J. and Liu, R. H., Mass spectrometry / forensic applications, in *Encyclopedia of Analytical Science*, 2nd ed., P. Worsfold, A. Townshend and C. Poole (Eds), Elsevier, Amsterdam, 2005, pp. 484–493.
6. Saferstein, R., Forensic applications of mass spectrometry, in *Forensic Science Handbook Vol.* **1**, 2nd ed., R. Saferstein (Ed.), Prentice Hall, Upper Saddle River, NJ, 2002, pp. 117–159.
7. Webb, K. S., The identification of drugs by mass spectrometry, in *The Analysis of Drugs of Abuse*, T. Gough (Ed.), John Wiley & Sons Ltd, Chichester, UK, 1991.
8. NIST/EPA/NIH Mass Spectral Library, 2005.
9. Szynkowska, M. I., Czerski, K., Rogowski, J., Paryjczak, T. and Parczewski, A., *Surf. Interface Anal.* **42**, 393–399 (2010).
10. Makinen, M., Nousialnen, M. and Sillanpaa, M., *Mass Spect. Rev.* **30**, 940–973 (2011).
11. Ifa, D. R., Jackson, A. U., Paglia G. and Cooks, R. G., *Anal. Bioanal. Chem.* **394**, 1995–2008 (2009).
12. Green, F. M., Salter, T. L., Stokes, P., Gilmore, I. S. and O'Connor, G., *Surface Int. Anal.* **42**, 347–357 (2010).
13. Yinon, J. (Ed.), *Forensic Applications of Mass Spectrometry*, CRC Press, Boca Raton, FL, 1995.
14. Meier-Augenstein, W., *Stable Isotope Forensics: An Introduction to the Forensic Application of Stable Isotope Analysis*, John Wiley & Sons Ltd, Chichester, UK, 2010.
15. Meier-Augenstein, W. and Liu, R. H., Forensic applications of isotope ratio mass spectrometry, in *Advances in Forensic Applications of Mass Spectrometry*, Y. Yinon (Ed.), CRC Press, Boca Raton, FL, 2004.
16. Benson, S., Lennard, C., Maynard, P. and Roux, C., *Forensic Sci. Int.* **157**, 1–22 (2006).
17. Brazier, J. L., Use of isotope ratios in forensic analysis, in *Forensic Applications of Mass Spectrometry*, J. Yinon (Ed.), CRC Press, Boca Raton, FL, 1995, pp. 259–289.
18. Ehleringer, J. R., Cerling T. E. and West, J. B., Forensic science applications of stable isotope ratio analysis, in *Forensic Analysis on the Cutting Edge: New Methods for Trace Evidence Analysis*, R. D. Blackledge (Ed.), John Wiley & Sons Ltd, Chichester, UK, 2007, pp. 399–422.
19. Daeid, N. N., Buchanan, H. A. S., Savage, K. A., Fraser J. G. and Cresswell, S. L., *Aust. J. Chem.* **63**, 3–7 (2010).

20. Karpas, Z., Ion mobility spectrometry in forensic science, in *Encyclopedia of Analytical Chemistry*, R. A. Meyers (Ed.), John Wiley & Sons, Inc., Hoboken, NJ, 2006.
21. Karpas, Z., Forensic applications of ion mobility spectrometry, *Forensic Sci. Rev.* **1**, 104–119 (1989).
22. Woodfin, R. L., Ion mobility spectrometry, in *Trace Chemical Sensings of Explosives*, R. L. Woodfin (Ed.), John Wiley & Sons Ltd, Chichester, UK, 2007.
23. Eiceman, G. A. and Schmidt, H., Advances in ion mobility spectrometry of explosives, in *Aspects of Explosives Detection*, M. Marshall and J. C. Oxley (Eds.), Elsevier, Amsterdam, 2009, pp. 171–202.
24. Ewing, R. G., Atkinson, D. A., Eiceman, G. A. and Ewing, G. J., *Talanta* **54**, 515–529 (2001).

Chapter 7

Separation Techniques

Learning Objectives

- To understand the origins of paper chromatography and how to obtain information from the technique for forensic samples.
- To apply paper chromatography to the study of documents.
- To understand the origins of thin layer chromatography and how to obtain information from the technique for forensic samples.
- To apply thin layer chromatography to the study of drugs, documents, fibres and explosives.
- To understand the origins of gas chromatography and how to obtain information from the technique for forensic samples.
- To apply gas chromatography to the study of drugs, toxicological samples, arson residues and explosives.
- To understand the origins of liquid chromatography and how to obtain information from the technique for forensic samples.
- To apply gas chromatography to the study of drugs, toxicological samples and fibres.
- To understand the origins of ion chromatography and how to obtain information from the technique for forensic samples.
- To apply ion chromatography to the study of explosives.
- To understand the origins of capillary electrophoresis and how to obtain information from the technique for forensic samples.
- To apply capillary electrophoresis to the study of drugs, toxicological samples, explosives and gunshot residues.

Forensic Analytical Techniques, First Edition. Barbara Stuart.
© 2013 John Wiley & Sons, Ltd. Published 2013 by John Wiley & Sons, Ltd.

7.1 Introduction

A particular challenge for forensic scientists is the characterization and quantification of substances of interest in complex matrices. For instance, biological specimens or explosive debris can require the separation of an analyte from a mixture prior to analysis. Chromatographic techniques have been widely used to deal with the separation of such samples. The general principle of chromatography is that one phase is held in place, while the other phase moves past (the mobile phase). The stationary phase can be solid particles or a liquid bonded to the inside of a capillary tube or onto the surface of solid particles packed in a column. Paper, thin layer, gas, liquid and ion chromatographies have all been used to deal with forensic samples. Separations may also be carried out under the influence of an electric field in capillaries using capillary electrophoresis techniques.

7.2 Paper Chromatography

Paper chromatography (PC) is the simplest chromatographic technique and, as the name implies, uses paper as the separation medium [1]. The paper, which is composed of cellulose, to which polar water molecules can be absorbed, acts as the stationary phase. The mobile phase consists of a less polar solvent, usually composed of an organic solvent and water. The paper is placed in a suitable solvent, and the solvent moves by capillary action through the paper and the sample.

7.2.1 Methods

A strip of filter paper is employed, and a sample in solution is placed as a spot near one edge of the paper. The sample components migrate at different characteristic rates that can be used to identify components. If the compounds are colourless, then reagents may be applied to reveal circular or elliptical spots. The method is simple, but there are several limitations. The sample cannot be volatile, the length of the migration path is limited and only qualitative analysis is possible. PC can also take some time and an experiment is often left for several hours to complete. Although PC has been superseded by other chromatographic techniques in terms of sensitivity, speed and the amount of information derived, this technique is straightforward and inexpensive so it can still be of assistance.

7.2.2 Interpretation

It is possible to identify compounds using PC by R_f values. The R_f value is:

$$R_f = \frac{\text{distance moved by compound}}{\text{distance moved by solvent front}} \qquad (7.1)$$

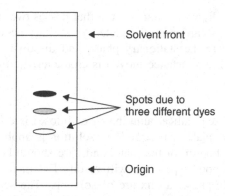

Figure 7.1 Paper chromatography of an ink.

The distance moved is measured relative to the point of application and is measured to the middle of the spot. Although R_f values are referenced, experimental conditions do affect the values, so a knowledge of the conditions used is required when comparing results.

7.2.3 Documents

PC is a simple test for the identification of inks [2]. Inks are commonly a mixture of dyes, each of which can be resolved using PC due to the difference in chemical structure. Figure 7.1 illustrates the appearance of an ink separated using PC. Ink samples can be extracted from a document for analysis using warm methanol before spotting onto a chromatography paper.

DQ 7.1

Why should a pencil be used for labelling spots on a chromatography paper?

Answer

A pencil is made of graphite, which will not migrate through the paper when exposed to the solvent and interfere with the interpretation of the results.

7.3 Thin Layer Chromatography

Thin layer chromatography (TLC) is a straightforward cost-effective chromatographic method for a range of forensic sample types [3–5]. In this technique, the stationary phase is coated on a plate and the sample mixture is spotted at one end.

The mobile phase is a liquid organic solvent that passes over the spot. Separation occurs as the solvent is carried up the plate by capillary action. Each compound in a mixture adheres to the stationary phase and dissolves in the solvent to a different extent. Thus, the distance moved is characteristic of a compound.

7.3.1 Methods

The stationary phase (e.g. silica, alumina or cellulose) is coated as a thin layer onto a glass or plastic plate. The sample in solution is applied as a spot using a capillary tube on the bottom of the plate, and care should be taken not to apply too much sample as poor separation will result. The plate is placed in a tank containing the mobile phase, a mixture of solvents. The separated components are usually detected by spraying the plate with a suitable chemical agent or observing under UV light (plates impregnated with fluorescing compounds are commercially available). High-performance TLC has been developed to improve the separation ability and the sampling time via the use of optimized plates.

DQ 7.2

Why should the origin spots be placed above the solvent level in the chamber?

Answer

If the spots are submerged in the solvent, they may be washed off the plate before separation can occur.

7.3.2 Interpretation

It is possible to identify compounds using TLC by determining R_f values, defined by Equation 7.1. Although R_f values have been collected for many compounds, the experimental conditions do affect the values, so a knowledge of the conditions used is required when comparing results. Figure 7.2 illustrates the appearance of a TLC plate.

SAQ 7.1

In a TLC experiment, the solvent front moves 8.5 cm and a compound in a sample being analysed moves 3.1 cm from the baseline. What is the R_f value in this case?

7.3.3 Drugs

TLC is a very common screening technique for the presence of drugs [3, 4, 6, 7]. Silica plates are the common stationary phase. Numerous solvent systems

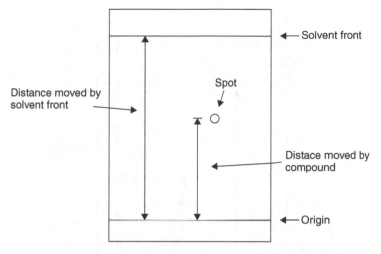

Figure 7.2 Separation of a TLC plate.

have been employed for drug systems, with ethyl acetate–methanol–ammonia a common system for basic drugs. Often plates impregnated with fluorescence compounds are used and examined using UV light, but various chemical reagents may be used to develop the TLC results. The acidified iodoplatinate reagent ($PtCl_4$ and KI in HCl solution) is a useful general developing reagent for drugs, demonstrating different colours for different classes of drugs.

SAQ 7.2

A developed TLC plate has been obtained to test for the presence of amphetamine and methamphetamine in an unidentified sample (shown in Figure 7.3). The predicted R_f values for amphetamine and methamphetamine in the development solvent used are 0.62 and 0.26, respectively. Is it possible using these TLC results to confirm the presence of amphetamine and methamphetamine in the unidentified sample?

7.3.4 Documents

TLC provides a useful tool for the characterization of inks used in questioned documents [3, 8, 9]. By comparing ink compositions by TLC, it is possible to establish the authenticity of a document. Ink samples can be obtained using a hypodermic needle or scalpel and are then dissolved in pyridine for ball-point inks or in ethanol or water for non-ballpoint inks. A variety of mobile phases can be used for the separation of the constituent dyes, including ethyl

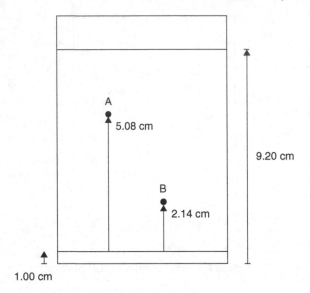

Figure 7.3 TLC plate for drug analysis (cf. SAQ 7.2).

acetate–ethanol–water, butanol–isopropanol–water and acetone–ethanol–water. After removal from this solvent and allowing it to dry, the plates can be observed under UV light to detect any fluorescence.

7.3.5 Fibres

TLC is a well-established tool for the separation of fibre dyes [3, 10–12]. Fibres of mm lengths are usually required for analysis, and the dye is solvent-extracted (pyridine–water is suitable for many samples) using a Pasteur pipette or capillary tube. Silica gel is typically used as the stationary phase. There are numerous solvent mixtures that can be used, and the choice will depend very much on the class of dye. It is advisable that multiple solvent systems are used for development to improve the quality of information obtained from the method. Some useful solvent systems for a range of dye classes are chloroform–methanol–ammonia–water and ethyl acetate–ethanol–water.

7.3.6 Explosives

The detection of explosives can be achieved by TLC using a modified Griess reagent as this reagent detects nitrates and most nitro-containing organic explosives [3]. Acetone or toluene are both used as solvents for the extraction of an explosive sample. A combination of mobile phases is required (e.g. ethyl acetate–petroleum ether and chloroform–methanol) to ensure separation of a range of commercial explosives. Visualization is carried out for a modified Griess reagent after hydrolysis with sodium hydroxide at 105°C. A pink-red colour indicates the presence of organic nitrites, and a red colour that fades to yellow indicates inorganic nitrites. If no colour is initially observed, then the addition of zinc powder will produce a red-pink colour of inorganic nitrates or organic nitro compounds are present.

7.4 Gas Chromatography

Gas chromatography (GC) involves the introduction of a gaseous or vaporized sample into a long column where the compounds with the sample are separated [13–16]. The components are flushed sequentially from the column to a detector and each component may be identified by measuring the retention time, the time taken to reach the detector.

7.4.1 Methods

Figure 7.4 illustrates a schematic diagram of a gas chromatograph. The volatile liquid or gaseous sample is injected via a septum into a heated port, and the vapour is carried through a temperature-controlled column by a carrier gas such as He or N_2. Gas liquid chromatography is a common form of GC and uses a nonvolatile liquid coated onto an inert solid support in a packed column. A new efficient approach that provides improved separations is capillary GC. A capillary column has the liquid (usually a siloxane-based polymer) as a thin layer on the inside of the usually silica thin column.

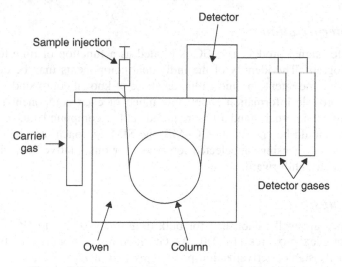

Figure 7.4 Schematic diagram of a gas chromatograph.

Various detectors are used in GC for forensic purposes. A *flame ionization detector* (FID) has a flame that ionizes the analyte, and a current results, producing the detector signal. A *thermal energy analyser* (TEA) is used to thermally decompose nitrogen-containing compounds. An *electron capture detector* (ECD) is a selective detector used to detect halides or oxygen-based molecules. A radioactive source is used to produce thermal electrons, which interact with electronegative analytes eluted from the column and produce the current detected.

The coupling of GC and MS is extremely valuable in a forensic laboratory as the spectrometer allows for the rapid identification of the separated components. The eluted gas is passed into a chamber to reduce to the low pressure required for MS analysis. The components are then ionized, and the spectrum recorded as described in Chapter 6.

Depending on the type of sample to be examined using GC, some sort of sample preparation method may be required (e.g. a drug in urine). A chemical process such as *hydrolysis* or *derivatization* may be required to produce an analyte suitable for analysis: certain compounds may be nonvolatile or thermally unstable. SPE methods have also been developed to purify samples prior to chromatographic analysis and allow the analyte to be adsorbed on a chosen surface and then eluted for analysis. *Solid-phase microextraction* (SPME) involves the extraction of analytes from gaseous or liquid samples without the need of a solvent – a coated silica fibre is used to absorb the analyte and is directly inserted into the instrument. *Headspace analysis* can be employed to examine volatile substances. In this approach, the volatile sample is sealed in an air-tight container and the vapours that build up inside the container are collected with a syringe for GC analysis. Pyrolysis of a sample is another tactic for dealing with complex samples, and the use of this technique in combination with GC is described in Chapter 8.

7.4.2 Interpretation

The detector signal produced in GC is plotted as a function of time to produce a chromatogram. The identity of the individual components may be determined by comparing the retention time with a database of known compounds. The peak areas can provide information about how much of each component is present. The use of an internal standard is required and a compound that elutes near the analyte should be chosen. In GC–MS, the SIM approach may be employed, which involves focusing on selected ion peaks for quantitative analysis to keep the process straightforward.

7.4.3 Drugs

GC methods are well established for bulk drug analysis due to GC's ability to resolve complex mixtures [14–18]. As GC requires compounds to be volatile and thermally stable, derivatization procedures may need to be carried out on a drug sample prior to analysis. There are various derivatization procedures used depending on the nature of the drug to be analysed. Some common derivatization reagents include *N*,*O*-bistrimethylsilyltrifluoroacetamide (BSTFA), *N*-methyl-*M*-trimethylsilyltrifluoroacetamide (MSTFA), trifluoroacetic anhydride (TFAA) and *N*,*O*-bistrimethylsilylic acid (BSA). GC with capillary columns and nonpolar phases with FID or MS detection are widely used.

SAQ 7.3

BSTFA is used as a derivatization reagent for THC prior to GC analysis. Draw the structure of the derivative produced. Why is derivatization required before analysis of THC?

GC can simply be used for qualitative analysis, where the presence of a particular drug is to be determined. Quantitative drug analysis is routinely carried out using GC and GC–MS. A quantitative approach to GC–MS analysis of drugs is SIM, particularly if the sample is complex. The area or height of the peak is measured and compared to an internal standard.

SAQ 7.4

GC is a good technique for the analysis of cocaine samples. A BSA-derivatized cocaine sample is analysed by GC–MS, with an internal standard (methyl docosanoate) employed in this experiment. A series of standard cocaine solutions was also analysed under the same experimental conditions. The cocaine peak areas obtained are listed in Table 7.1. Determine the concentration of cocaine in the sample. Comment on the quality of the calibration data in this experiment.

GC is also useful for drug profiling. The impurities and adulterants present in illicit drugs can be identified and quantified to link a sample to a source or to

Table 7.1 Calibration data for GC–MS analysis of a cocaine sample (cf. SAQ 7.4)

Cocaine concentration (mg ml^{-1})	Internal standard (arbitrary units)	Cocaine peak area (arbitrary units)
0.127	9.21×10^6	5.33×10^4
0.290	8.99×10^6	7.64×10^4
0.472	9.01×10^6	1.22×10^5
0.613	8.03×10^6	1.34×10^5
0.837	8.56×10^6	1.86×10^5
1.014	9.56×10^6	2.46×10^5
sample	9.28×10^6	1.77×10^5

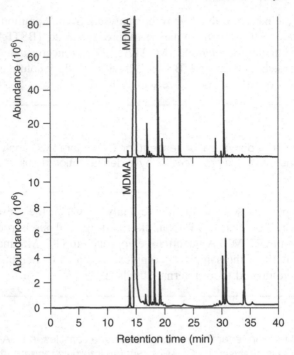

Figure 7.5 Gas chromatograms for MDMA produced by reductive amination (upper) and the Leukart method (lower). Reproduced with permission from G. Zadora and D. Zuba, 'Gas chromatography in forensic science', in *Encyclopedia of Analytical Chemistry*, ed. R.A. Meyer, Wiley, New York, 2009.

the method of production. Figure 7.5 illustrates the chromatograms obtained for MDMA samples prepared by reductive amination and the Leukart method. There are clear differences in the impurity profiles based on the method of production of MDMA that can be identified using GC.

7.4.4 Toxicology

GC techniques are valuable in forensic toxicology, where drugs are contained in body fluids or tissues. GC is established as the most widely used technique for the analysis of alcohol in body fluids, especially the blood alcohol content (BAC) [15, 16, 19]. GC enables ethanol to be distinguished from other alcohols and can detect to concentrations on the order of $0.001 \, gl^{-1}$. Figure 7.6 provides an example of a chromatogram for a breath alcohol sample. In addition to ethanol, other volatiles are readily separated and identified. Direct injection and headspace analysis are the most common procedures for BAC determination. The direct injection method involves the addition of an internal standard such as propanol. Headspace analysis has the advantage of minimizing the possibility of contamination of the syringe

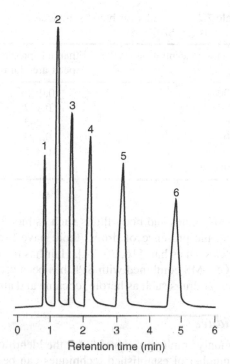

Figure 7.6 Gas chromatography of a breath alcohol sample. Reproduced with permission from Vidmantas *Ralsys, J. Chem. Ed.* 60, 62, 12, 1050–1051 (1985).

or column. According to Henry's law, the ethanol concentration in the headspace is proportional to the concentration of ethanol in the solution. Thus, the ethanol concentration in the sample may be determined by measurement of the peak area (or height) of the headspace sample. A FID is commonly used as the detector because it is sensitive to alcohol while being insensitive to water.

SAQ 7.5

GC–FID is used to determine the alcohol concentration of a blood sample. A series of ethanol standard solutions is prepared, the details of which are listed in Table 7.2. GC analysis was carried out on the standards, as well as on an unknown blood sample. An internal standard of 1 ml *n*-propanol diluted to 50 ml in water was used in the experiment. Peaks in a resulting chromatogram were observed at 0.60 and 0.39 min for the *n*-propanol and ethanol, respectively. The area ratios of the ethanol peak divided by the *n*-propanol peak are also listed in Table 7.2 for the standards and the sample. Determine the BAC for the unknown sample.

Table 7.2 GC results for blood sample and standards (cf. SAQ 7.5)

Ethanol concentration (mg ml^{-1})	Ethanol/n-propanol peak area ratio
0.200	0.0816
0.400	0.162
0.600	0.243
0.800	0.324
1.00	0.405
unknown	0.292

Although analyses of tissues and body fluids such as blood and urine are more common for detecting the presence of drugs, there have been developments in the analysis of samples such as hair [16, 20, 21]. Hair has the potential to identify long-term drug use. GC–MS combined with SPE has been used for the identification and quantification of drugs such as heroin, cocaine and amphetamines in hair.

7.4.5 *Arson Residues*

GC is the most commonly employed approach for the identification of accelerants [14–16, 21–26]. A number of established techniques can be used to collect and analyse ignitable liquid residues (ILRs). Traditional techniques have involved distillation and solvent extraction procedures, but these have been largely replaced by headspace sampling techniques, which extract and concentrate the debris volatiles onto an absorbent material for analysis [27–30]. The headspace techniques mostly use charcoal followed by carbon disulfide elution and may be passive or static (diffusion of volatiles to the adsorbent surface) or dynamic (where the headspace is forced into the adsorbent tube). The introduction of SPME has been a successful development for the sampling of accelerants [31]. The passive headspace–SPME method involves extraction of the volatiles by absorption into a polymer such as polydimethylsiloxane (PDMS) coated onto a fibre that can be directly inserted into the GC and thermally desorbed onto the column.

The most widely used techniques for ILR analysis are GC–FID and GC–MS. Capillary GC columns with silicone-based stationary phases are suitable for the hydrocarbons in petroleum-based liquids. Figure 7.7 illustrates a chromatogram obtained for a gasoline sample collected from a carpet using headspace sampling by SPME. GC–FID is an established approach, but now tends to be utilized as a screening method as GC–MS is more effective for the identification of ILRs. Pattern recognition techniques may be carried out on the chromatograms produced for ILRs. ILRs can be classified according to the constituent n-alkane carbon numbers: light (C4–C9), medium (C8–C13) and heavy (C9–C20+). The mass spectral data obtained using GC–MS enables the well-established characteristic

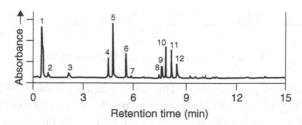

Figure 7.7 GC chromatogram of a gasoline sample collected from a carpet using SPME headspace sampling: (1) toluene; (2) 3-methylheptane; (3) octane; (4) ethylbenzene; (5) 1,3-dimethylbenzene and 1,4-dimethylbenzene; (6) 1,2-dimethylbenzene; (7) nonane; (8) propylbenzene; (9) 1-ethyl-(2 or 3)-methylbenzene; (10) 1,2,5-trimethylbenzene; (11) 1-ethyl-2-methylbenzene; (12) 1,2,4-trimethylbenzene. Reproduced with permission from the *Journal of Forensic Sciences*, **48**, 130–134, J.A. Lloyd and P.L. Edmiston, *J. Forensic Sci*. Copyright 2003 ASTM International, 100 Barr Harbor Drive, West Conshohocken, PA 19428.

ions of ILRs to be identified. The compound types present in ignitable liquids consist of normal alkanes, branched alkanes, cycloalkanes, aromatics, polynuclear aromatics and oxygenates. Target compounds analysis, which uses the identification of specific compounds for different classes of ignitable liquid, aid in the identification of a sample.

SAQ 7.6

If kerosene is classified as a heavy ignitable liquid, what type of information is expected from a GC analysis of kerosene?

7.4.6 Explosives

In the past, GC methods were unsuitable for explosives analysis because of the thermal instability and polarity of many explosive materials. However, capillary GC and capillary GC–MS methods have been developed to detect explosives, particularly for the analysis of post-blast debris [14–16]. Sensitive detectors such as the ECD and the TEA can be employed to analyse to picogram sensitivity.

7.5 Liquid Chromatography

Liquid chromatography (LC) is a chromatographic technique that has a liquid mobile phase. LC proves useful when the compounds under investigation are not thermally stable or are not sufficiently volatile for GC analysis [32–34].

7.5.1 Methods

High-resolution separation of compounds is achieved using *high-performance liquid chromatography* (HPLC), where a solvent is forced through a column of

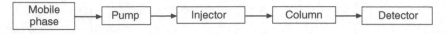

Figure 7.8 Layout of an HPLC instrument.

small particles (5–10 μm diameter). The main components of an HPLC apparatus are shown in Figure 7.8. HPLC requires a mobile phase that dissolves the analyte in question. A pump delivers a flow of solvent to the column containing the stationary phase. Isocratic elution is carried out with a single or constant solvent, but if this does not provide a suitable elution, then gradient elution can be employed. Gradient elution involves adding increasing amounts of a second solvent to the first solvent to produce a continuous gradient.

The separation of compounds in a sample analysed using LC is a result of the different interactions of the compounds with the stationary phase. In normal-phase LC, a polar silica stationary phase is used with a nonpolar solvent such as hexane. This method is useful for the analysis of nonpolar compounds. Reverse phase LC has a nonpolar stationary phase (often chemically modified silica) and a polar solvent, such as water, can be used.

The separated compounds that emerge from the LC column reach a detector. The most common detector type is a UV-visible absorbance detector often using a *diode array detector* (DAD). This type of detector measures the UV-visible spectrum of the solution, and nanogram quantities of analyte can be measured. *Fluorescence detectors* may also be useful for particular forensic applications. A mass spectrometer can also act as a detector, and the development of *LC–MS* has improved the information obtained. When the mobile phase reaches the mass spectrometer in LC–MS, the solvent is removed (now feasible due to electrospray ionization (ESI) technology) and the analytes pass into the spectrometer for analysis.

7.5.2 Interpretation

The data output using LC is similar to that of GC. That is, a chromatogram showing peaks due to the various components as a function of retention time is obtained. The retention times may possibly be used for identification purposes, but the use of LC–MS provides more confirmatory information from the mass spectra of the components.

Where a UV detector is used, quantitative analysis can be carried out, assuming that Beer's law is applicable. The linearity of the absorbance recorded with concentration must be established before this approach is applied. A calibration plot over the required concentration range should be prepared.

7.5.3 Drugs and Toxicology

LC techniques have emerged as a viable alternative to GC for the identification of drug-based evidence [13, 17, 35–38]. LC has the advantage of allowing nonvolatile analytes to be examined in a nondestructive manner, thus enabling the sample to be recovered for further analysis. Derivatization is not generally necessary. If dealing with toxicological specimens, an extraction process is required to separate the sample from the matrix. LC was less often applied until recent times due to the large volume of solvents required. However, the introduction of ESI and APCI techniques has led to the expansion of LC for forensic drug analysis. LC–MS has grown as a regular technique for drug analysis and toxicology. UV-visible and fluorescence methods may be used for drug detection, and the choice of detection depends on the analyte structure.

A variety of stationary phases and solvent systems are used to examine drugs, and normal- and reverse-phase LC are used. Normal phase often utilizes silica as a stationary phase, and for reverse phase hydrocarbons are commonly available. The mobile phase can be combinations of solvents such as water, methanol, acetonitrile and tetrahydrofuran. Some drugs can become charged in a buffer and may interact with the stationary phase causing distortion in the output, but the addition of an acid for acidic drugs or a base for basic drugs produces better results (known as *ion suppression chromatography*). An *ion-pairing method*, which involves inducing a charge on the solute and countering with an opposite ion to produce a neutral species, may also be used to deal with charged drugs.

The separation of a mixture of opiates is illustrated in the chromatogram in Figure 7.9.

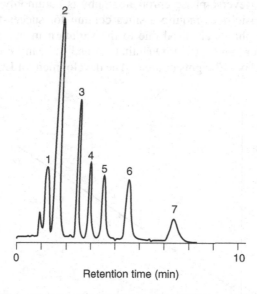

Retention time (min)

Figure 7.9 HPLC separation of opiates: (1) noscapine; (2) papaverine; (3) acetylcodeine; (4) diamorphine; (5) 6-monoacetylmorphine; (6) codeine; (7) morphine. Reproduced with permission from M.J. Bogusz, 'Liquid chromatography/mass spectrometry in forensic toxicology', in Advances in Forensic Applications of Mass Spectrometry, ed. Y. Yinon, CRC Press, Boca Raton, 2004.

SAQ 7.7

Explain the order of elution of acetylcodeine and morphine observed in the chromatogram shown in Figure 7.9.

SAQ 7.8

An amphetamine sample of unknown concentration is prepared in acidified methanol, and a series of standard amphetamine solutions is prepared under the same conditions. A calibration plot showing the UV absorbance at 259 nm versus amphetamine free base concentration is shown in Figure 7.10. The regression equation resulting is $y = 0.0299x + 0.00001$. Determine the concentration of amphetamine in a sample that produces an absorbance of 0.0070. Is it possible using this calibration data to determine the concentration of an amphetamine sample that produces an absorbance of 0.070?

7.5.4 Fibres

Fibre dyes may be classified using LC techniques [39, 40]. As with TLC, extraction of the dye is required prior to LC analysis, with pyridine–water being a common solvent. Reverse-phase chromatography is commonly used to separate many dyes, but basic dyes require a silica column for successful separation. A variety of mobile phases are used due to the variation in dye chemistry, and a UV-visible detector can be used. Quantitative analysis can be carried out using retention times and peak heights or areas. The development of LC–MS techniques

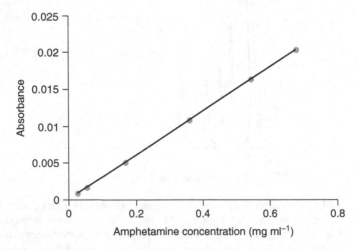

Figure 7.10 LC calibration graph for amphetamine (cf. SAQ 7.8).

with ESI has improved the discrimination ability of LC. With MS, identification of dyes is feasible and ESI enables nonvolatile dyes to be converted into a gas phase.

7.6 Ion Chromatography

Ion chromatography (IC) is based on an attraction between solute ions and the charged sites bound to a stationary phase, and is used to separate charged compounds [41, 42]. The stationary phase is usually an ion exchange resin that carries charged functional groups that interact with oppositely charged groups of the compound to be retained. That is, a positively charged anion exchanger interacts with anions, while a negatively charged cation exchange interacts with cations. The bound compounds can be eluted from the column by gradient elution or by elution with a change in pH or salt concentrations.

7.6.1 Methods

Ion exchange resins are amorphous particles of materials such as copolymer resins composed of styrene and divinylbenzene, the relative amounts of which result in different degrees of cross-linking in the resin. The aromatic groups may be modified to contain SO_3 groups to produce a cation exchange resin, or modified with NR_3^+ groups to produce an anion exchange resin. Conductivity measurement provides a method of detecting ions in solution. Any electrolytes that might interfere with the analysis can be removed using suppressed ion chromatography prior to detection by electrical conductivity. In suppressed ion–anion chromatography, the solution passes through a suppressor in which the cations are replaced by H^+ to convert the eluent to H_2O. Suppressed ion–cation chromatography has a suppressor which replaces the anion with OH^-. UV detection is also possible for certain systems.

7.6.2 Interpretation

As with GC and LC, the detector response is plotted as a function of retention time. The peaks can be used to identify the species of ion present when compared to results obtained for standards. The peak area is also proportional to the concentration of the ion species.

7.6.3 Explosives

IC is a suitable technique for the analysis of inorganic pre- and post-blast explosives at the ppm level [43–45]. Anions, including nitrate, chlorate and perchlorate, and cations, such as ammonium, potassium and sodium, are readily detected. A range of columns, mobile phases and detectors can be used to separate the various analytes observed in explosives. Dilute water extracts of samples are used, with a

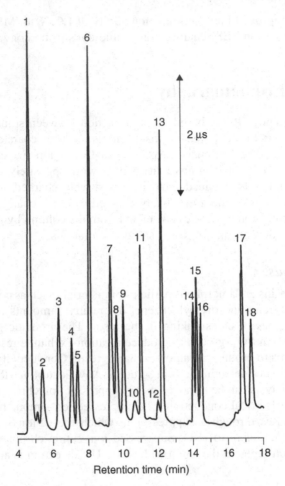

Figure 7.11 Ion chromatogram of anion standards for improvised explosive devices: (1) fluoride; (2) acetate; (3) formate; (4) chlorite; (5) bromated; (6) chloride; (7) nitrite; (8) cyanate; (9) chlorate; (10) benzoate; (11) nitrate; (12) carbonate; (13) sulfate; (14) phosphate; (15) thiosulfate; (16) chromate; (17) thiocyanate; (18) perchlorate. Reproduced with permission from C. Johns, R.A. Shellie, O.G. Potter, J.W. O'Reilly, J.P. Hutchinson, R.M. Guijit, M.C. Breadmore, E.F. Hilder, G.W. Dicinoski and P.R. Haddad, 'Identification of homemade inorganic explosives by ion chromatographic analysis of post-blast residues', *J. Chromatog*. A 1182, 205–214 (2008).

high-purity water source required to avoid contamination. Figure 7.11 illustrates an ion chromatogram of the standards of anions expected in improvised explosive devices. A well-established approach to explosives analysis is the use of both IC and capillary electrophoresis (described in Section 7.7). These two techniques have complementary selectivities for the components found in explosives.

7.7 Capillary Electrophoresis

Electrophoresis involves the migration of ions in solution under the influence of an electric field. When a potential is applied across a buffer containing molecules of interest using electrodes, the ions of the sample migrate towards one of the electrodes. The rate of migration depends on the charge and the size of the molecule. Capillary electrophoresis (CE) uses a capillary to allow high electric fields to be applied, resulting in better resolution and a shorter analytical time than traditional electrophoresis [33, 46–48].

7.7.1 Methods

Figure 7.12 illustrates the layout of a CE instrument. A voltage in the range of 10–30 kV is applied across a buffer-filled capillary. Capillaries are typically fused silica and require only very small volumes. The migration of ions is detected typically by a UV-visible, fluorescence or conductometric detector.

A number of modes of CE are available, and the types more commonly used for forensic samples are *capillary zone electrophoresis* (CZE) and *micellar electrokinetic capillary chromatography* (MECC or MEKC). CZE is carried out in a continuous buffer, and the separation is based on differences in the electrophoretic mobilities. MEKC was developed to separate neutral species, which are difficult to separate using CZE. In MEKC an ionic surfactant is added to the buffer so micelles can interact with neutral analytes and facilitate migration to an electrode. Portable CE is emerging via 'lab-on-a-chip' technology, and interfacing CE with MS will expand the application of this technique in forensic science.

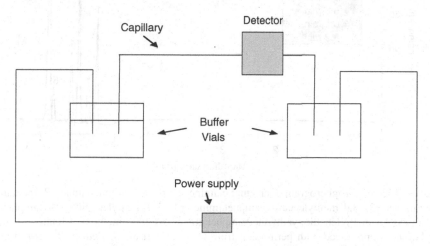

Figure 7.12 Layout of a CE instrument.

7.7.2 Interpretation

In CE, the detector output is used to produce an electropherogram. The electropherogram shows the peaks representing the separated compounds as a function of migration time. Qualitative analysis can be carried out using the migration times compared to standards. The peak areas can be used for quantitative analysis with the use of standard solutions.

7.7.3 Drugs and Toxicology

CE is used for forensic drug analysis, with MECC and CZE being the most widely used techniques, with a variety of detection methods employed [46, 49–51]. CE is used for the analysis of drug seizures to identify the principal constituents. This enables the chemical profiling of illicit drugs. CE is also used in forensic toxicology, with specimens in blood and urine being the most widely investigated. Hair can also be tested for drugs using CE methods.

One of the most common applications of CE to drug analysis is to the examination of amphetamine-type stimulants, providing a suitable alternative to GC and LC, and no sample derivatization is required [52]. The separation of the different enantiomers present can be accomplished by the addition of what is known as a chiral selector to the electrolyte. For example, cyclodextrin or various derivatives can be used to separate the enantiomers of amphetamine and methamphetamine.

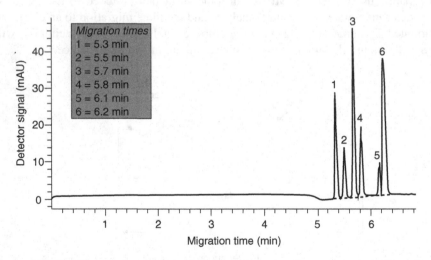

Figure 7.13 Electropherogram of an amphetamine sample: (1) amphetamine; (2) methamphetamine; (3) 3,4-methylenedioxyamphetamine; (4) 3,4-methylenedioxyethylamphetamine; (5) 3,4-methylenedioxyethylmethamphetamine; (6) 2,5-dimethoxy-4-methyl-phenethylamine. Reproduced with permission from G. Boatto et al., *J. Pharm. Biomed. Anal.* **29**, 1073–1080 (2002) Elsevier.

Figure 7.13 shows an electropherogram of an amphetamine sample and illustrates the separation of the main enantiomers.

DQ 7.3

An illicit drug sample believed to contain a mixture of neutral, acidic and basic drugs is to be analysed using CE. Would CZE or MEKC be the better choice of CE separation technique for this sample?

Answer

As this sample contains neutral drugs, MEKC is the better choice. MEKC enables neutral, basic and acidic drugs to be simultaneously analysed.

7.7.4 Explosives and Gunshot Residues

CE has developed as a complementary technique to IC for the analysis of explosive and gunshot residues [43, 46, 50]. IC shows good reproducibility and sensitivity, and CE offers efficiency and short run times. Both inorganic and organic explosive materials can be identified using this approach. The minaturization and development of portable devices make CE a particularly useful technique for the detection of explosives and GSR. A range of CE separation methods have been applied to this type of evidence. The challenge when dealing with the separation of inorganic cations is the similar electrophoretic mobilities exhibited by these species. This problem can be solved by adding a weak complexing agent to the electrolyte.

7.8 Summary

Separation techniques, including chromatography and electrophoresis, provide a means of separating the mixtures often encountered in forensic evidence. The simplest separation technique is paper chromatography, and this can be used as a quick comparison test for document analysis. TLC can be used to confirm the presence of a range of compounds. Despite its relative lack of specificity and resolution, TLC is still a valuable tool in a forensic laboratory as it can be used as a simple and inexpensive preliminary test. GC, often combined with MS, is very widely used as a tool for the qualitative and quantitative analysis of drugs, toxicological samples, arson residues and explosives. LC is also an important separation technique for forensic samples and is particularly useful for samples that are not suitable for GC analysis. Drugs, toxicological samples, fibres and explosives can be examined using LC techniques. IC is a more specialized chromatographic technique for forensic samples, but does have value when analysing explosives. CE is also used to separate the components of complex mixtures, including drugs, toxicological samples, explosives and GSR.

References

1. Gasparic, J. and Churacek, J., *Laboratory Handbook of Paper and Thin Layer Chromatography*, Ellis Horwood, New York, 1978.
2. Aginsky, V., Document analysis/analytical methods, in *Encyclopedia of Forensic Sciences*, J. Siegel, G. Knupfer and P. Saukko (Eds), Academic Press, New York, 2000, pp. 566–570.
3. Ardrey, R. E., Thin layer chromatography, in *Encyclopedia of Analytical Science*, 2nd ed., P. Worsfold, A. Townshend and C. F. Poole (Eds), Elsevier, Amsterdam, 2005, pp. 481–485.
4. Moffat, A. C., Osselton, M. D., Widdop, B. and Galichet, L. Y., *Clarke's Analysis of Drugs and Poisons*, 3rd ed., Pharmaceutical Press, London, 2004.
5. Reich, E. and Blatter, A., Thin layer chromatography/overview, in *Encyclopedia of Analytical Science*, 2nd ed., P. Worsfold, A. Townshend and C. Poole (Eds), Elsevier, Amsterdam, 2005, pp. 57–66.
6. Ojanopera, I., Forensic toxicology: thin layer (planar) chromatography, in *Encyclopedia of Separation Science*, Academic Press, New York, 2000, pp. 2879–2885.
7. Cole, M. D. and Caddy, B., *The Analysis of Drugs of Abuse: An Instruction Manual*, Ellis Horwood, New York, 1995.
8. Pagano, L. W., Surrency, M. J. and Cantu, A. A., Inks: forensic analysis by thin layer (planar) chromatography, in *Encyclopedia of Separation Science*, Academic Press, New York, 2000, pp. 3101–3109.
9. ASTM Standard E1422, *Standard Guide for Test Methods for Forensic Writing Ink Comparison*, American Society for Testing and Materials, West Conshohocken, PA, 2005.
10. Wiggins, K. G., Thin layer chromatographic analysis for fibre dyes, in *Forensic Examination of Fibres*, 2nd ed., J. Robertson and M. Grieve (Eds), CRC Press, New York, 1999.
11. Gaudette, B. D., The forensic aspects of textiles fibre examination, in *Forensic Science Handbook Vol. 2*, R. Saferstein (Ed.), Prentice Hall, Englewood Cliffs, NJ, 1998, pp. 209–272.
12. Goodpaster, J. V. and Liszewski, E. A., *Anal. Bioanal. Chem.* **394**, 2009–2018 (2009).
13. Levine, B. (Ed.), *Principles of Forensic Toxicology*, 2nd ed., AACC Press, Washington, DC, 2006.
14. Zadora, G. and Zuba, D., Gas chromatography in forensic science, in *Encyclopedia of Analytical Chemistry*, R. A. Meyer (Ed.), John Wiley & Sons, Inc., New York, 2009.
15. Tebbett, I., *Gas Chromatography in Forensic Science*, Ellis Horwood, Chichester, 1992.
16. Brettell, T. A., Forensic science applications of gas chromatography, in *Modern Practice of Gas Chromatography*, 4th ed., R. L. Grob and E. F. Barry (Eds), John Wiley & Sons, Ltd, Chichester, 2004, pp. 883–967.
17. Gough, T. A., *The Analysis of Drugs of Abuse*, John Wiley & Sons, Ltd, Chichester, 1991.
18. Maurer, H. H., Screening for drugs in body fluids by GC–MS, in *Advances in Forensic Applications of Mass Spectrometry*, Y. Yinon (Ed.), CRC Press, Boca Raton, 2004.
19. Zabzdyr, J. L. and Lillard, S. J., *J. Chem. Ed.* **78**, 1225–1227 (2001).
20. Dasgupta, A., Drugs of abuse, analysis of, in *Encyclopedia of Analytical Chemistry*, R. A. Meyer (Ed.), John Wiley & Sons, Inc., New York, 2008.
21. Yinon, J. (Ed.), *Forensic Applications of Mass Spectrometry*, CRC Press, Boca Raton, 1995.
22. Baron, M., Forensic sciences/arson residues, in *Encyclopedia of Analytical Science*, 2nd ed., P. Worsfold, A. Townshend and C. Poole (Eds), Elsevier, Amsterdam, 2005, pp. 365–372.
23. Neumann, H., Gas chromatography/forensic applications, in *Encyclopedia of Analytical Science*, 2nd ed., P. Worsfold, A. Townshend and C. Poole (Eds), Elsevier, Amsterdam, 2005, pp. 139–146.

24. Dolan, J., *Anal. Bioanal. Chem.* **376**, 1168–1171 (2003).
25. Sandercock, P. M. L., *Forensic Sci. Int.* **176**, 93–110 (2008).
26. ASTM Standard E1618, *Standard Test Method for Ignitable Liquid Residues in Extracts from Fire Debris Samples by Gas Chromatography–Mass Spectrometry*, American Society for Testing and Materials, West Conshohocken, PA, 2011.
27. ASTM Standard E1412, *Standard Practice for Separation of Ignitable Liquid Residues from Fire Debris Samples by Passive Headspace Concentration with Activated Charcoal*, American Society for Testing and Materials, West Conshohocken, PA, 2007.
28. ASTM Standard E1413, *Standard Practice for Separation and Concentration of Ignitable Liquid Residues from Fire Debris Samples by Dynamic Headspace Concentration*, American Society for Testing and Materials, West Conshohocken, PA, 2007.
29. ASTM Standard E1386, *Standard Practice for Separation of Ignitable Liquid Residues from Fire Debris Samples by Solvent Extraction*, American Society for Testing and Materials, West Conshohocken, PA, 2010.
30. ASTM E1388, *Standard Practice for Sampling of Headspace Vapours from Fire Debris Samples*, American Society for Testing and Materials, West Conshohocken, PA, 2005.
31. ASTM E2154, *Standard Practice for Separation and Concentration of Ignitable Liquid Residues from Fire Debris Samples by Passive Headspace Concentration with Solid Space Microextraction (SPME)*, American Society for Testing and Materials, West Conshohocken, PA, 2008.
32. Bayne, S. and Carlin, M., *Forensic Applications of High Performance Liquid Chromatography*, CRC Press, Boca Raton, 2010.
33. Northrup, D., Forensic applications of high performance liquid chromatography and capillary electrophoresis, in *Forensic Science Handbook Vol. 1*, 2nd ed., R. Saferstein (Ed.), Prentice Hall, Upper Saddle River, NJ, 2002, pp. 41–116.
34. Ardrey, R. E., *Liquid Chromatography – Mass Spectrometry: An Introduction*, John Wiley & Sons, Ltd, Chichester, 2003.
35. Wood, M., Laloup, M., Samyn, N., del Mar Ramirez Fernandez, M., de Bruijn, E. A., Maes, R. A. A. and de Boeck, G., *J. Chromatog. A* **1130**, 3–15 (2006).
36. Bogusz, M. J., Liquid chromatography/mass spectrometry in forensic toxicology, in *Advances in Forensic Applications of Mass Spectrometry*, Y. Yinon (Ed.), CRC Press, Boca Raton, 2004.
37. van Boexlaer, J. F., Clauwaert, K. M., Lambert, W. E., Deforce, D. L., van den Eeckhout, E. G. and de Leenhear, A. P., *Mass Spect. Rev.* **19**, 165–214 (2000).
38. Cole, M. D., *The Analysis of Controlled Substances*, John Wiley & Sons, Ltd, Chichester, 2003.
39. Watson, N., Forensic sciences/fibres, in *Encyclopedia of Analytical Science*, 2nd ed., P. Worsfold, A. Townshend and C. Poole (Eds), Elsevier, Amsterdam, 2005, pp. 406–414.
40. Griffin, R. and Peers, J., Other methods of colour analysis 12.1 High performance liquid chromatography, in *Forensic Examination of Fibres*, 2nd ed., J. Robertson and M. Grieve (Eds), Taylor and Francis, London, 1999.
41. Weiss, J. and Weiss, T., *Handbook of Ion Chromatography*, Wiley-VCH, Berlin, 2004.
42. Yashin, Y. I., Yashin, A. Y. and Walton, H. F., Ion exchange/ion chromatography instrumentation, in *Encyclopedia of Analytical Science*, 2nd ed., P. Worsfold, A. Townshend and C. Poole (Eds), Elsevier, Amsterdam, 2005, pp. 453–460.
43. Hutchinson, J. P., Johns, C., Dicinoski, G. W. and Haddad, P. R., Identification of improvised inorganic explosives devices by analysis of post-blast residues using ion chromatography and capillary electrophoresis, in *Encyclopedia of Analytical Chemistry*, R. A. Meyer (Ed.), John Wiley & Sons, Inc., New York, 2008.
44. Dicinoski, G. W., Shellie, R. A. and Haddad, P. R., *Anal. Lett.* **39**, 639–657 (2006).
45. Johns, C., Shellie, R. A., Potter, O. G., O'Reilly, J. W., Hutchinson, J. P., Guijit, R. M., Breadmore, M. C., Hilder, E. F., Dicinoski, G. W. and Haddad, P. R., *J. Chromatog. A* **1182**, 205–214 (2008).

46. Tagliaro, F. and Pascali, V. L., Capillary electrophoresis in forensic science, in *Encyclopedia of Forensic Sciences*, J. Siegel, G. Knupfer and P. Saukko (Eds), Academic Press, New York, 2000, pp. 135–146.
47. Sadecka, J., Forensic sciences/capillary electrophoresis, in *Encyclopedia of Separation Science*, Academic Press, New York, 2000, pp. 2862.
48. Tagliaro, F., Manetto, G., Crivellente, F. and Smith, F. P., *Forensic Sci. Int.* **92**, 75–88 (1998).
49. Anastos, N., Barnett, N. W. and Lewis, S. W., *Talanta* **67**, 269–279 (2005).
50. Cruces-Blanco, C., Gamiz-Gracia, L. and Garcia-Campana, A. A., *Trends Anal. Chem.* **26**, 215–226 (2007).
51. Cruces-Blanco, C. and Garcia Campana, A. M., *Trends Anal. Chem.* **31**, 85–95 (2012).
52. Chinaka, S., Iio, R., Takayama, N., Kodama, S. and Hayakawa, K., *J. Health Sci.* **52**, 649–654 (2006).

Chapter 8
Thermal Analysis

8.1 Introduction

Thermal analysis involves the measurement of physical and chemical changes that a material undergoes as it is heated. These changes can include decomposition processes, the release or absorption of energy or a weight loss or gain. Such changes occur at temperatures that are characteristic to a particular material. Thermal methods that are useful to the forensic scientist are pyrolysis techniques, differential scanning calorimetry, differential thermal analysis and thermogravimetric analysis.

Forensic Analytical Techniques, First Edition. Barbara Stuart.
© 2013 John Wiley & Sons, Ltd. Published 2013 by John Wiley & Sons, Ltd.

8.2 Pyrolysis Techniques

Pyrolysis involves heating a substance at high temperatures in an inert atmosphere. The process produces molecular fragments that are characteristic of the starting material. In order to identify the pyrolysis products, a pyrolyser is coupled with a gas chromatograph, a mass spectrometer or a combination of both apparatus [1–6].

8.2.1 Methods

Three common types of pyrolyser are available in commercial instruments. Furnace pyrolysers involve placing the sample in a preheated reactor. In a Curie point (or inductive-heating) pyrolyser, the sample is placed in a ferromagnetic wire within a radiofrequency field. A resistive-heating pyrolyser uses a resistively heated platinum wire to heat a sample. A microgram quantity of sample is loaded into the pyrolyser and is rapidly heated, usually to temperatures in the range of $600 - 800°C$. Reproducible experimental conditions are an important consideration when using pyrolysis techniques in order to obtain reliable data because competing secondary reactions may occur if varying temperature conditions are employed.

Following pyrolysis, the products are sent to a gas chromatograph or a mass spectrometer, depending on the choice of system. In *pyrolysis gas chromatography* (Py–GC), the pyrolysis products are separated and identified with a gas chromatograph. The resulting chromatogram produced is known as a pyrogram. *Pyrolysis–capillary GC* provides an even more sensitive approach: the use of a capillary column instead of a packed column results in improved resolution. The eluate from the GC may be further analysed with a mass spectrometer. *Pyrolysis GC–MS* (Py–GC–MS) provides the ability to identify the separated components. It is also feasible to analyse forensic samples using *pyrolysis mass spectrometry* (Py–MS), although this is used less commonly than the Py–GC techniques.

8.2.2 Interpretation

The type of fragments that result from the thermal decomposition of molecules depends on the molecular structure of the starting material and the thermal conditions. There can be a number of possible degradation pathways. For polymers, some typical pyrolysis reactions are depolymerization (where the polymer returns to its monomer form), side group scission (where the groups attached to the backbone are broken and the backbone becomes unsaturated) and random chain scission (where the polymer backbone is randomly broken). Table 8.1 lists the characteristic species that may be identified as a result of pyrolysis of some common polymers.

A standard approach to using pyrolysis data is the comparison of the pyrogram of an unknown with a library of standard pyrograms. Pyrograms are, for the

Table 8.1 Pyrolysis products of polymers

Polymer	Main pyrolysis products
acrylics	ethyl acrylate, methyl methacylate, butyl acrylate, 2-ethylhexylacrylate, 2-ethylhexyl alcohol, isooctane
butadiene	vinylcyclohexene
epoxy resins	phenol, iopropenyl phenol, bisphenol A
polyesters	benzene, benzoic acid, biphenyl terephthalate, vinyl terephthalate
polyethylene	hydrocarbons, n-hexane, n-pentane
polyisoprene	isoprene, dipentene, dimethylvinylcyclohexene
poly(methyl methacrylate)	methyl methacrylate
polypropylene	hydrocarbons, 2,4-dimethylpeptane
polystyrene	styrene, benzene, toluene, ethylbenzene
poly(vinyl acetate)	acetic acid
poly(vinyl chloride)	hydrogen chloride, benzene, toluene, chlorobenzenes
silicones	siloxanes

most part, complex, so it is unusual to use all the peaks in a pyrogram for comparison purposes, and generally a selection of the most intense peaks is made for comparison. Care must be exercised when dealing with retention times, but problems can be avoided with the addition of a standard: the retention time of the standard can be used to correct the retention times of the unknown sample.

8.2.3 Paint

Py–GC techniques are used to identify and differentiate the organic binders of forensic paint specimens [3–9]. Information about additives and impurities may also be obtained using this approach. In order to enhance the information provided by pyrolysis techniques, paints can be derivatized, and the process is referred to as thermally assisted hydrolysis methylation (THM). A mixture of the sample and a tetraalkylammonium hydroxide (commonly tetramethylammonium hydroxide (TMAH)) is heated, and alkyl derivatives are produced that provide additional discriminating peaks in the pyrogram.

SAQ 8.1

The pyrogram of the acrylic layer of a vehicle paint is shown in Figure 8.1. Given that the main numbered peaks shown have been identified, what is the copolymer composition of the layer?

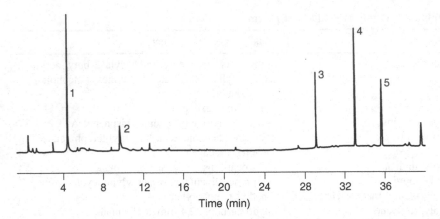

Figure 8.1 Pyrogram of an acrylic paint: (1) methyl methacrylate; (2) methacrylic acid; (3) dibutyl phthalate; (4) butyl cyclohexyl phthalate; (5) butyl benzyl phthalate. Reproduced with permission from J.M. Challinor, 'Examination of forensic evidence', in Applied Pyrolysis Handbook, 2nd ed., Taylor and Francis, London, 2007, pp 175–199.

8.2.4 Fibres

Both synthetic and natural fibres can be identified by using pyrolysis methods [5, 6, 10]. Py–GC and Py–GC–MS are particularly useful as the decomposition products of the different classes can be discriminated with microgram quantities of sample while using these techniques. THM can also be used for fibres to improve the sensitivity where required.

SAQ 8.2

What pyrolysis products would be expected to appear in a pyrogram of nylon 6, 6 fibre?

8.2.5 Polymers

Pyrolysis techniques are widely used to characterize other polymer-based materials. Py–GC can be used to determine the composition of rubber samples collected from crime scenes, such as those left by vehicle accidents or shoe soles [1, 5, 6]. The technique permits discrimination of the common rubber types as they show distinct decomposition products. Py–GC can also be used to determine the composition of rubber blends, thus adding to the discriminating ability of the technique. Quantitative analysis can be carried out by measuring the ratios of the peaks due to pyrolysis products in the resulting pyrogram. For example, the relative amounts of isoprene, butadiene and styrene in vehicle types vary between brands. By measuring the peaks due to vinylcyclohexene (a pyrolysis product of butadiene), styrene and dipentene (a pyrolysis product of isoprene), the composition can be determined. Multivariate analysis can also be used to handle complex data.

The pyrolysis products of adhesives can be used to classify an unknown specimen [5, 11, 12]. Table 8.1 contains the main pyrolysis products of some common adhesive components, and these may be used to identify the class of adhesive used. Further discrimination may be obtained by exploiting the differences in additives, such as plasticizers, between commercial adhesives.

8.2.6 Documents

The use of Py–GC in document analysis focuses on the analysis of inks and toners [2, 5]. For inks, the dye and solvents can be characterized. For toners, the polymer binders can be determined. For sampling, the region of interest can be scraped with a scalpel, or a small section can be removed.

8.3 Differential Scanning Calorimetry and Differential Thermal Analysis

The thermal methods of differential scanning calorimetry (DSC) and differential thermal analysis (DTA) are widely used to characterize the physical and chemical properties of a variety of materials [13–19]. DSC is a technique that records the energy necessary to establish a zero temperature difference between a sample and a reference material as a function of temperature or time. The two specimens are subjected to identical temperature conditions in an environment heated or cooled at a controlled rate. DTA involves measuring the difference in temperature between the sample and the reference material as a function of temperature or time.

8.3.1 Methods

The layout of the apparatus used for DSC and DTA experiments is illustrated in Figure 8.2. A small sample of the order of milligram quantities is contained in a crucible (often an aluminium pan), then placed in a furnace. Powders, film or fibres can be examined. The reference material is often alumina (Al_2O_3). The sample chamber can be purged with a gas to control the atmosphere. The furnace is electrically heated and heating rates up to $100°C\,min^{-1}$ can be used, but a normal rate is $10°C\,min^{-1}$. Temperatures below room temperature can be measured by using a coolant such as liquid N_2.

8.3.2 Interpretation

DSC curves are plotted with heat flow commonly as a function of temperature at a constant rate of heating. A shift in the baseline results from the change in

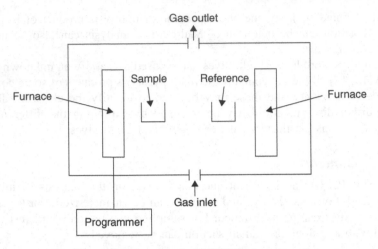

Figure 8.2 Layout of a DSC/DTA instrument.

heat capacity of the sample, and the basic equation used for DSC is:

$$\Delta T = \frac{qC_p}{K} \qquad (8.1)$$

where ΔT is the difference in temperature between the reference material and the sample, q is the heating rate, C_p is the heat capacity of the sample and K is a calibration factor for the instrument. DSC can also be used for measuring enthalpy in transitions. The peak area between the curve and the baseline is proportional to the enthalpy change (ΔH) in the sample. ΔH can be determined from the area of the curve peak (A) by using:

$$\Delta H\, m = K\, A \qquad (8.2)$$

where m is the mass of the sample. A DTA curve is usually a plot of the temperature difference versus temperature.

DSC and DTA are established techniques for characterizing the thermal properties of polymers, and the application of these techniques to forensic samples is often for polymer-based evidence. Transitions due to specific thermal processes can be identified by using DSC or DTA. The *glass transition temperature* (T_g) is the temperature at which a polymer ceases to be glassy and becomes rubbery. The T_g of a crystallizable polymer can be detected by using DSC as an endothermic shift from the baseline and is observed due to an increase in the heat capacity due to the increased molecular motions in the polymer. The T_g value is affected by a number of factors, including the nature of substituent groups attached to the polymer backbone, the copolymer structure, the type of bonding between chains, cross-linking, the molecular weight and the presence of plasticizers, as

Table 8.2 Melting and softening temperatures of some common polymers

Polymer	Melting or softening temperature ($^\circ$C)
PS	70–115
PVC	75–90 (softens)
PE	120–135
PMMA	120–160
PP	165–180
nylon 11	180–190
nylon 6	210–220
nylon 6,6	250–260
PET	250–260

well as the heating and cooling rates used in a DSC run. Although the beginning of the transition is often defined as the T_g, the standard procedure involves the use of regression lines and the point of inflection is commonly used as the T_g value. Polymers also exhibit a *crystalline melting temperature* (T_m), the temperature range over which crystalline polymers melt, and an endothermic peak is observed in the DSC curve. Such a peak enables the melting point and the enthalpy of melting to be determined using DSC. Table 8.2 lists the T_m values for some common polymers. For crystallizable polymers, the *crystallization temperature* (T_c) is the temperature at which ordering and the production of crystalline regions occur, and an exothermic peak is observed in a DSC curve.

8.3.3 Polymers

DSC and DTA can be used to characterize the polymers commonly used in packaging materials collected as evidence [14, 17, 20]. The DSC traces of low-density polyethylene (LDPE) and high-density polyethylene (HDPE) are shown in Figure 8.3. The differences in the melting temperatures for the different structures are clear in the DSC results: an endotherm correlating to a T_m of 115°C for LDPE is observed, while a T_m of 130°C is determined for the HDPE sample. The differences in thermal behaviour are due to the different degree of chain branching in the two types of PE: LDPE has a greater degree of branching than HDPE. The areas of the endotherms can be used to determine the percentage crystallinity of PE, thus providing a further means of discriminating specimens.

8.3.4 Fibres

DSC and DTA are established techniques for the study of synthetic and natural fibres [16, 21]. These techniques have the advantage of measuring high melting temperatures compared to the use of a hot stage. The T_g and T_m values provide a means of identifying and discriminating fibres. The requirements for milligram

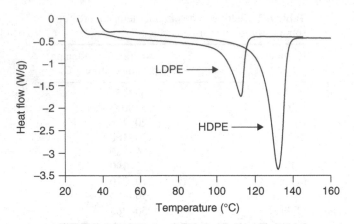

Figure 8.3 DSC traces of LDPE and HDPE.

quantities of fibre can be a limitation as only microgram amounts of fibre may be available from a crime scene. However, higher sensitivity instrumentation is emerging that enables samples of the order of micrograms to be accurately measured, and this may lead to the expansion of the technique to forensic fibre analysis.

SAQ 8.3

Figure 8.4 illustrates the DSC trace obtained for a nylon fibre. Identify the type of nylon used to produce this fibre.

8.4 Thermogravimetric Analysis

Thermogravimetric analysis (TGA) is a thermal method that involves measurement of the mass loss of a material, mostly as a function of temperature [14, 15, 22, 23]. TGA is used to quantify the mass changes in a material associated with various transitions or degradation processes. Different substances show unique patterns of such processes at specific temperatures, and thus TGA provides characteristic mass loss data for a given material.

8.4.1 Methods

In TGA, a sample is placed in a furnace while being suspended from the arm of a sensitive balance. The change in sample mass is recorded while the sample is maintained at the required temperature or subjected to a programmed heating. The instrument layout of TGA is illustrated in Figure 8.5. Depending on

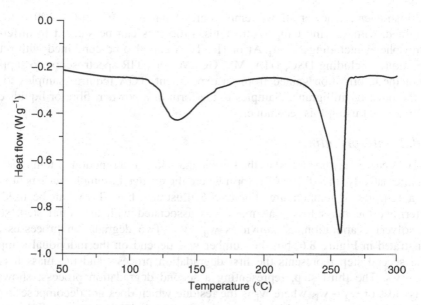

Figure 8.4 DSC traces of a nylon (cf. SAQ 8.3).

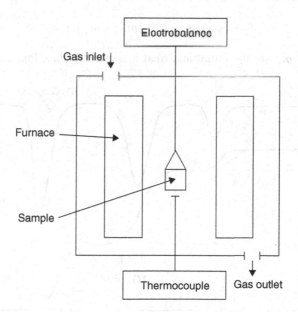

Figure 8.5 Layout of a TGA instrument.

the instrument, a heater allows temperatures from $-196°C$ up to $2400°C$ to be obtained, with varying temperature rates. Processes can be studied in different atmospheres including N_2, O_2, Ar or He. TGA can also be combined with other techniques including DSC, DTA, MS, GC–MS or FTIR spectroscopy to supplement the information gained from an experiment. TGA requires samples to be of the order of milligrams. Samples in the form of a powder, fibre or liquids can be examined using this technique.

8.4.2 Interpretation

A TGA curve may be plotted as the sample mass loss as a function of temperature, or, alternatively, in a differential form where the change in sample mass is plotted as a function of temperature. Figure 8.6 illustrates how TGA can be used to determine the mass loss. The mass loss associated with an initial step, such as solvent evaporation, is shown as $w_0 - w_1$. Two degradation processes are illustrated in Figure 8.6, but the number will depend on the individual sample. The second step represents the first degradation process, and the mass loss is $w_1 - w_2$. The third step, representing a second degradation process, shows a mass loss of $w_2 - w_f$, where w_f is the residue which does not decompose in the temperature range covered by the experiment. The derivative curve (DTG) shows a peak associated with each separate step representing the maximum rate of mass loss. The percentage of mass loss can be determined using:

$$\% \text{ mass loss} = 100 \left(m_i - m_f\right) / m_i \qquad (8.3)$$

where m_i and m_f are the initial and final masses in a mass loss step.

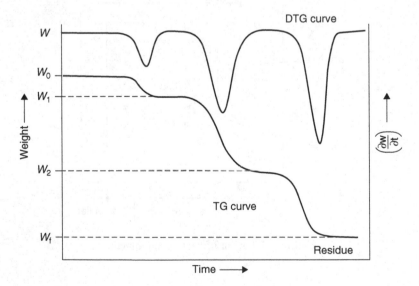

Figure 8.6 TGA mass loss steps.

8.4.3 Polymers

TGA enables polymers to be characterized based on the fact that the thermal degradation of polymeric materials depends very much on composition, a result of the range of degradation processes that can be observed for polymers [24–26]. For example, in the DTG data collected for a PVC sample in a N_2 atmosphere, peaks in the ranges of 250–350°C and 400–500°C can be observed. These peaks are associated with the elimination of HCl and the carbon backbone scission reactions, respectively, identified in the degradation of PVC.

One of the most useful aspects of TGA is the identification and quantification of additives. Polymer additives such as fillers and plasticizers can be detected based on their individual decomposition processes, and the amount determined based on mass loss calculations. For example, $CaCO_3$ will decompose to CaO, or $Al_2O_3 - 3H_2O$ can eliminate water to produce Al_2O_3. Figure 8.7 shows the TG data obtained for an elastomer containing carbon black. The experiment is initially carried out in an inert N_2 atmosphere to degrade other organic components, and this is followed by heating in O_2 to oxidize the carbon black components.

SAQ 8.4

A polymer containing $CaCO_3$ as a filler is analysed using TGA. The TG data indicate that there is a mass change from 249.7 to 191.5 mg between 600 and 900°C, which is associated with the decomposition of $CaCO_3$ to CO_2. Determine the percentage of $CaCO_3$ in the sample.

8.4.4 Explosives and Arson Residues

TGA provides a useful method for the examination of polymer evidence retrieved from fires or explosions [25]. The properties of a polymer can be significantly affected by high temperatures, so TGA provides a means of characterizing

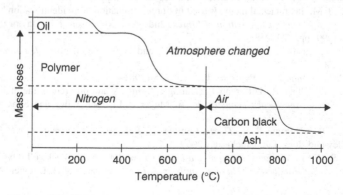

Figure 8.7 TGA data for an elastomer containing carbon black.

polymer evidence. For instance, if the retrieved material contains a carbon chain incorporated into plastic evidence collected from an explosion site, it is still possible to identify a polymer.

8.5 Summary

Thermal techniques that involve monitoring changes to materials as they are heated provide a means of characterizing forensic materials based on the different responses to heat by specimen type. The thermal methods that are most relevant to forensic scientists are pyrolysis GC and GC–MS, DSC, DTA and TGA. These techniques are particularly useful for the study of polymer-based evidence including fibres, paint, rubber, adhesives, toners and packaging, as well as for explosives and arson residues.

References

1. Blackledge, R. D., Pyrolysis gas chromatography in forensic science, in *Encyclopedia of Analytical Science*, 2nd ed., Elsevier, Amsterdam, 2005, pp. 4524–4536.
2. Sobeih, K. L., Baron, M. and Gonzalez-Rodriquez, J., *J. Chromatography A* **1186**, 51–66 (2008).
3. DeForest, P. R., Crim, D., Tebbett, I. R. and Larsen, A. K., Pyrolysis gas chromatography in forensic science, in *Gas Chromatography in Forensic Science*, I. Tebbett (Ed.), Prentice Hall, Old Tappan, NJ, 1994, pp. 165–185.
4. Wheals, B. B., *J. Anal. Appl. Pyrolysis* **2**, 277–292 (1980–1981).
5. Challinor, J. M., Examination of forensic evidence, in *Applied Pyrolysis Handbook*, 2nd ed., Taylor and Francis, London, 2007, pp. 175–199.
6. Challinor, J. M., *Forensic Sci. Int.* **21**, 269–285 (1983).
7. Challinor, J. M., Pyrolysis techniques for the characterisation and discrimination of paint, in *Forensic Examination of Glass and Paint*, B. Caddy (Ed.), Taylor and Francis, London, 2001.
8. ASTM Standard E1610, *Standard Guide for Forensic Paint Analysis and Comparison,* American Society for Testing and Materials, West Conshohocken, PA, 2002.
9. Buzzini, P. and Stoecklein, W., Forensic sciences/paints, varnishes and lacquers, in *Encyclopedia of Analytical Science*, 2nd ed., Elsevier, Amsterdam, 2005, pp. 453–464.
10. Challinor, J. M., Instrumental methods used in fibre examination: fibre identification by pyrolysis techniques, in *Forensic Examination of Fibres*, 2nd ed., J. Robertson (Ed.), Taylor and Francis, London, 1999, pp. 223–238.
11. Huttunen, J., Austin, C., Dawson, M., Roux, C. and Robertson, J., *Aust. J. Forensic Sci.* **39**, 93–106 (2007).
12. Curry, C. J., *J. Anal. Appl. Pyrolysis* **11**, 213–225 (1987).
13. Riga, A. and Collins, R., Differential scanning calorimetry and differential thermal analysis in *Encyclopedia of Analytical Chemistry*, R. A. Meyer (Ed.), John Wiley and Sons, Chichester, 2000.
14. Menczel, J. D. and Prime, R. B., *Thermal Analysis of Polymers: Fundamentals and Applications*, John Wiley and Sons, Ltd, Chichester, 2009.
15. Gabbott, P. (Ed.), *Principles and Applications of Thermal Analysis*, Blackwell, Oxford, 2007.
16. Cheng, S. Z. D., *Handbook of Thermal Analysis and Calorimetry*, Vol. 3, Elsevier, Amsterdam, 2002.

17. ASTM Standard D3418, *Standard Test Method for Transition Temperatures and Enthalpies of Fusion and Crystallization of Polymers by Differential Scanning Calorimetry*, American Society for Testing and Materials, West Conshohocken, PA, 2008.
18. ASTM Standard E1356, *Standard Test Method for Assignment of the Glass Transition Temperatures by Differential Scanning Calorimetry*, American Society for Testing and Materials, West Conshohocken, PA, 2008.
19. ASTM Standard E794, *Standard Test Method for Melting and Crystallization Temperatures by Thermal Analysis*, American Society for Testing and Materials, West Conshohocken, PA, 2006.
20. ASTM Standard D7426, *Standard Test Method for Assignment of the DSC Procedure for Determining T_g of a Polymer or Elastomeric Compounds*, American Society for Testing and Materials, West Conshohocken, PA, 2008.
21. ASTM Standard D7138, *Standard Test Method to Determine Melting Temperature of Synthetic Fibres*, American Society for Testing and Materials, West Conshohocken, PA, 2008.
22. Dunn, J. G., Thermogravimetry, in *Encyclopedia of Analytical Chemistry*, R. A. Meyer (Ed.), John Wiley & Sons, Ltd, Chichester, 2000.
23. Dollimore, D. and Phang, P., Simultaneous techniques in thermal analysis, in *Encyclopedia of Analytical Chemistry*, R. A. Meyer (Ed.), John Wiley and Sons, Ltd, Chichester, 2000.
24. Stuart, B. H., *Polymer Analysis*, John Wiley & Sons, Ltd, Chichester, 2002.
25. Ihms, E. C. and Brinkman, D. W., *J. Forensic Sci*. **49**, 505–510, 2004.
26. ASTM Standard D6370, *Standard Test Method for Rubber – Compositional Analysis by Thermogravimetry (TGA)*, American Society for Testing and Materials, West Conshohocken, PA, 1999.

Responses to Self-Assessment Questions

Chapter 1

Response 1.1

(a) The most common type of polymer used in the manufacture of bags is PE.

(b) The tape components of adhesive tape may be manufactured from polymers including cellulose, PP and PVC. Polymers based on butadiene, isoprene, acrylates or methacrylates may be identified in the adhesive component.

(c) SAN and SBR copolymers are commonly used in the manufacture of car tyres.

Response 1.2

Disperse dyes are commonly used to colour polyester fibres. For this type of dye, van der Waals forces and hydrogen bonding are responsible for the interaction between fibre and dye.

Response 1.3

(a) Toughened glass is used to produce car windscreens, and a windscreen may also be laminated.

Forensic Analytical Techniques, First Edition. Barbara Stuart.
© 2013 John Wiley & Sons, Ltd. Published 2013 by John Wiley & Sons, Ltd.

(b) A blow-moulding process is used to produce glass for containers such as bottles.

Response 1.4

The explosives that contain a C-N-NO$_2$ functional group are HMX and RDX.

Response 1.5

Figure 1.6 illustrates the structures of morphine and heroin (diamorphine), and it can be seen that the OH groups are acetylated. Inspection of the codeine structure shows an OH structure that will be acetylated to produce acetylcodeine, the structure of which is:

Chapter 2

Response 2.1

The purple colour produced by the Ehrlich test provides evidence of the presence of LSD – this is recognized as a fast screening test for LSD. The results of the Mandelin test also provide evidence of the presence of LSD, with a grey colour produced in this test. Thus, the results of these two tests can aid in the identification of the drug type.

Response 2.2

The TMB test is more sensitive than the Kastle–Meyer test, so the TMB test may be detecting lower concentrations of blood. The possibility of a false positive result in the TMB should also be considered, as the contamination by other substances (e.g. cosmetics) can contribute to a positive result.

Response 2.3

A green-grey colour resulting from a dithiooxamide test indicates the presence of copper. Copper is a common component of bullet jackets, and particulate matter from the jacket is produced during firing.

Response 2.4

The density of the mixture can be determined using Equation (2.2). If, say, a 100 cm^3 volume is prepared, then the density of the resulting solution is:

$$P = \frac{2.89 \text{ g cm}^{-3} \times 50 \text{ cm}^3 \times 1.00 \text{ g cm}^{-3} \times 50 \text{ cm}^3}{100 \text{ cm}^3}$$

$$= 1.95 \text{ g cm}^{-3}$$

Response 2.5

The specimen is observed to float in solutions containing 60–100% bromoform, while it sinks in solutions containing 40% or less bromoform. Therefore, the density of the specimen must have a density equating to that of a bromoform–bromobenzene solution in the range of 40–60% bromoform. Thus, the density of the specimen is of the order of 2.1–2.3 g cm^{-3}.

Response 2.6

In a saturated NaCl solution, a glass specimen sinks because glasses have densities of the order of 2 g cm^{-3}. Plastics have densities of the order of 1 g cm^{-3}, so they will float in the NaCl solution. More information about the type of glass used in the headlamp can be obtained by carrying out a density gradient experiment using bromoform–bromobenzene solutions or using a balance.

Response 2.7

(a) A polymer banknote has a nonporous surface, so the recommended sequence of fingermark detection is:
optical cyanoacrylate fuming VMD luminescent dye.

(b) A paper banknote has a porous surface, so the recommended sequence of fingermark detection is:
optical DFO ninhydrin metal salt treatment PD.

Chapter 3

Response 3.1

A birefringence value of 0.150 potentially corresponds to a Nomex, PET or PBT fibre based on the data listed in Table 3.1. Inspection of the n_{11} and $n_{\perp}$ values would enable the fibre to be narrowed to Nomex or to PET or PBT, as Nomex has higher RI values than PET or PBT. Determination of the melting temperature would further discriminate the fibre as each of the possibilities has quite distinct melting temperatures.

Response 3.2

As the RI values fall into the range typically found for headlight glass, this glass is very likely to have come from a broken headlight.

Response 3.3

Consultation with Table 3.3 shows that RIs of 1.54 and 1.55 correspond to the hexagonal mineral quartz. As quartz can form a hexagonal structure and shows two RIs, it is uniaxial.

Response 3.4

Mica is likely to be susceptible to damage by a preparation method such as microtoming, which involves cutting across the specimen. The use of ion milling would be a better approach as damage to the specimen would be minimized prior to examination.

Response 3.5

As there is the potential for aluminium to be present in the elemental composition of GSR, it is best to avoid the use of this type of adhesive tape. Carbon tape is a better choice as this will not interfere with the X-ray lines of interest.

Response 3.6

Consultation with Table 3.4 indicates that a yellow pigment that contains Cr, Pb and S is most likely to be chrome yellow.

Response 3.7

Fe is present in a number of inorganic pigments, including Prussian blue, ochres, red oxides and magnetite. As only Prussian blue in this group imparts a blue colour, the presence of Fe strongly indicates that this was the pigment used to colour the analysed fibre.

Response 3.8

An identification of the free base form of cocaine or the hydrochloride salt may be separately identified as they will produce different crystalline structures. Diluents will also be present so that a match with XRD patterns of sugars or other stimulants should be expected.

Chapter 4

Response 4.1

(a) A powder lends itself to examination using diffuse reflectance, but if such an accessory is not available then the sample can be mixed with KBr and a disc produced for transmission spectroscopy.

(b) A single textile fibre is very small in size so a microspectroscopy technique is required. A DAC is often used to press a fibre to a thickness suitable for transmission.

(c) Paint may be collected from a flat surface using an abrasive sampling pad, and a thin powdered layer is deposited on the surface. The pad can then be examined using a diffuse reflectance accessory.

Response 4.2

A comparison with the spectra of common binders in Figures 4.3–4.5 reveals that the binder is most likely to be PVA. There are characteristic bands due to $C=O$ stretching and C—O stretching near 1700 and $1000 \, cm^{-1}$, respectively. Note that because this spectrum is recorded in reflectance mode, the relative intensities of the bands appear different compared to the reference spectrum recorded in transmission mode (Figure 4.5). Additional bands are also observed due to the presence of a $BaSO_4$ filler, such as the sharp band observed near $3700 \, cm^{-1}$. No bands due to the pigment are observed in the spectrum as, in this case, cobalt blue is the pigment and does not show mid-infrared bands.

Response 4.3

A comparison with the spectrum of polypropylene in Figure 4.13 shows a good match with the spectrum in Figure 4.17. The spectrum shows characteristic bands associated with C—H and C—C bonds, with no evidence of other functional groups evident.

Response 4.4

The spectrum shows the characteristic bands associated with poly(*cis*-isoprene), confirmed by a comparison with the reference spectrum shown in Figure 4.21. Characteristic bands due to aliphatic C—H stretching in the $3000 - 2800 \, cm^{-1}$ region, $C=C$ stretching near $1670 \, cm^{-1}$ and C—H bending in the $1500 - 1300 \, cm^{-1}$ region correlate with an isoprene structure.

Response 4.5

The spectrum in Figure 4.24 shows indicative bands of styrene and acrylate. Polystyrene has C—H stretching bands above $3000 \, cm^{-1}$, indicating an aromatic group, and characteristic bands seen in the $2000-1700 \, cm^{-1}$ and $760-700 \, cm^{-1}$

regions. The appearance of a strong band at $1730\,\text{cm}^{-1}$ is a clear indication of a carbonyl group. Although deciphering the fingerprint region is complex, comparison with the reference spectra in Figure 4.4 shows a good match for PMMA. Thus, the toner contains a styrene–acrylate copolymer resin, a commonly used polymer system in toners.

Response 4.6

The principal difference between the pseudoephedrine and methamphetamine structures is the presence of an OH group in pseudoephedrine and the absence of this type of group in methamphetamine due to the reduction reaction used to produce this amphetamine. The OH group shows a distinct band near $3600\,\text{cm}^{-1}$, which will be observed in the pseudoephedrine spectrum and absent in the methamphetamine spectrum.

Response 4.7

The important structural difference between heroin and morphine is the presence of ester groups in heroin. This means that an additional band at $1740\,\text{cm}^{-1}$ due to the C=O stretching of the ester bond in heroin will be observed. Morphine does not contain an ester group, so this band will be absent in the spectrum.

Response 4.8

Raman microscopy is an appropriate choice of approach for a small specimen. If a laser source of 785 nm is employed, the form of TiO_2 can be characterized. Anatase will show Raman bands at 640, 520 and $400\,\text{cm}^{-1}$, while rutile will show bands at 1450, 1000, 980, 615 and $450\,\text{cm}^{-1}$.

Response 4.9

The broad band is due to fluorescing compound, most likely an additive present in the paper. The use of a longer wavelength laser source may improve the quality of the spectrum by eliminating the fluorescence peak and revealing the underlying sample peaks of interest.

Response 4.10

A comparison of the Semtex and RDX spectra shows that the strong bands present in RDX, a triplet at 1211, 1270 and $1304\,\text{cm}^{-1}$ and a triplet at 1347, 1381 and $1432\,\text{cm}^{-1}$, are absent in the spectrum of the Semtex sample. There is a good match with the PETN spectrum so the Semtex contains only PETN.

Response 4.11

An examination of the UV absorption maxima recorded in acidic solution for amphetamine and methamphetamine will show that there are similar bands observed for these compounds. The band maxima are observed at 251, 257 and 263 nm for amphetamine, and methamphetamine shows maxima at 252, 257 and 263 nm. In basic solution, these compounds show UV maxima at the same values of 258 and 267 nm. Given the similarity of the band maxima, it is not recommended that UV–vis spectroscopy be used to differentiate the amphetamine and methamphetamine.

Response 4.12

A calibration graph of absorbance versus concentration is first plotted and is shown in Figure SAQ 4.12. A linear calibration graph is observed. An absorbance of 0.432 corresponds to a concentration of 15.5 μg ml^{-1}, so the concentration of the unknown cocaine solution is 15.5 μg ml^{-1}. The assumption made for this calculation is that the band at 233 nm is free of interference from other bands in this region of the spectrum. If there is any overlap with adjacent peaks, the band height of interest will be distorted and the absorbance readings will be incorrect. Consequently, error will be introduced into the concentration value determined.

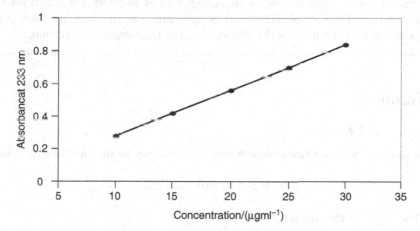

Figure SAQ 4.12 Calibration graph for cocaine solutions.

Response 4.13

If it is assumed that there is a linear relationship between the fluorescence intensity and the analyte concentration in this experiment, then:

$$\text{concentration of quinine in urine} = \frac{\text{urine extract intensity}}{\text{quinine reference intensity}} \times 1 \, \text{mg ml}^{-1}$$

$$= (0.76/0.92) \times 1 \, \text{mg ml}^{-1}$$

$$= 0.83 \, \text{mg ml}^{-1}$$

The experiment described relies on only one standard solution. A more reliable approach would be the preparation of a series of standard solutions and recording the fluorescence spectra for these in order to prepare a calibration curve.

Response 4.14

The peaks near 7 ppm in the spectrum are assigned to the benzene group present in amphetamine. The split peaks near 3, 2.5 and 1 ppm are due to the varying C—H environments within the molecule, and these may be assigned based on the ratios of the peak areas. Thus, the peaks above 3 ppm are due to C—H. Those centred at about 2.5 ppm are due to the methylene groups, while the more intense peaks above 1 ppm are due to the presence of the methyl group. The remaining broader peak near 2.5 ppm can be attributed to the NH_2 group.

Response 4.15

Consultation with Figure 1.10 indicates that the triplet at 1.0 ppm may due to the terminal methyl group. The quartet at 2.5 ppm can be attributed to the methylene groups and the phenyl group is responsible forth resonance peak at 7.4 ppm. As the peak in the maleic acid NMR spectrum that is used for quantitative analysis is observed at 6.2 ppm, the use of this compound as an internal standard for the quantitative analysis of phenobarbital is appropriate as the standard peak does not overlap with the main peaks observed in the phenobarbital spectrum.

Chapter 5

Response 5.1

The modified solution has an absorbance of 0.462, so the increase in absorbance is:

$$0.462 - 0.210 = 0.252$$

The Mg^{2+} of the added solution is:

$$0.010 \times 10^{-3}1 \times 1.00\,gl^{-1} = 1.00 \times 10^{-5}\,g = 10.0\,\mu g$$

This means that an absorbance of 0.252 corresponds to a concentration of 10.0 μgl^{-1}. Thus, an absorbance of 0.210 corresponds to a concentration of:

$$(0.210/0.252) \times 10.0\,\mu g\,l^{-1} = 8.33\,\mu g\,l^{-1}.$$

Response 5.2

A series of standard lead solutions should be prepared and analysed using GFAAS. The recorded absorbance values are then plotted versus concentration to produce a linear calibration graph. The blood sample should be analysed using identical experimental conditions, and the absorbance value recorded.

The lead concentration for the sample can then be determined using the calibration graph.

Response 5.3

According to the results in Table 5.1, the lead isotope ratios determined for the retrieved particle show a closer correspondence with those determined for bullet 1. Bullet 2 shows ratios with higher values. Thus, the particle can potentially be linked to bullet 1 and weapon 1 is predicted to be the one used to fire the fatal bullet. No error is quoted here, but of course the reproducibility of the results should always be tested.

Response 5.4

The Ca/Fe ratios of sheet and container glass vary, and this ratio can be measured using XRF. The specimen showing the higher Ca/Fe ratio can be attributed to the container glass, while the lower ratio enables the other specimen to be recognized as a sheet glass.

Response 5.5

The presence of Pb and Cr in a yellow paint is a strong indication of the presence of a chrome yellow pigment, which is composed of $PbCrO_4$. The presence of Ti indicates that TiO_2 was used as an extender in the paint.

Chapter 6

Response 6.1

The molecular mass of 6-monoacetylmorphine is $327\,g\,mol^{-1}$, so its parent ion would be observed at an *m/z* ratio of 328.

Response 6.2

The molecular masses of caffeine and paracetamol are calculated to be 194 and $151\,g\,mol^{-1}$, respectively. This indicates that molecular ion peaks in the spectrum should be observed at an *m/z* ratio of 195 for caffeine and 152 for paracetamol. Thus, in the sample examined, the presence of caffeine is indicated.

Response 6.3

A base peak at *m/z* = 227 indicates a structure with a molecular mass of 228. In Figure 1.5, this mass corresponds to only to the structure of TNT.

Response 6.4

Using Equation 6.1:

$$\delta\% = \frac{\left(R_{\text{sample}} - R_{\text{standard}}\right) \times 1000}{R_{\text{standard}}}$$

$$= (0.01085 - 0.01124) \times 1000 / 0.01124$$

$$= -34.7$$

Response 6.5

Inspection of the IMS spectra in Figure 6.5 reveals that only one common explosive type, pentaerythritol tetranitrate (PETN), shows distinct drift times in the 7–8 ms range under the conditions used. Thus, this specimen most likely contains PETN.

Chapter 7

Response 7.1

The R_f value is determined by Equation (7.1), so for this experiment:

$$R_f = \frac{3.1\,\text{cm}}{8.5\,\text{cm}} = 0.37$$

Response 7.2

The R_f values calculated for the two spots observed on the plate shown in Figure 7.3 are:

A:

$$R_f = \frac{\text{distance moved by compound}}{\text{distance moved by solvent front}} = \frac{5.08\,\text{cm}}{8.20\,\text{cm}} = 0.62$$

B:

$$R_f = \frac{2.14\,\text{cm}}{8.20\,\text{cm}} = 0.26$$

The spot at A has an R_f value corresponding to amphetamine and B corresponds to methamphetamine, so both of these compounds can be identified in this sample.

Response 7.3

The THC derivative produced has the structure shown in Figure SAQ 7.3. The presence of the OH group in THC needs to be modified for a successful separation via GC – this group can sorb onto the column and provide poor results.

Figure SAQ 7.3 BSTFA derivative of THC.

Response 7.4

In order to determine the cocaine concentration in the sample, a calibration plot of the peak area ratio of the internal standard and the cocaine peak versus the cocaine concentration needs to be prepared. The relevant peak area ratios are listed in Table SAQ 7.4, and a plot of these values versus the standard cocaine concentration is shown in Figure SAQ 7.4. The sample shows a peak area ratio of 0.0191, which corresponds to a cocaine concentration of 0.724 mg ml^{-1}. The plot passes close to zero (0.0025), but the data point at a concentration of 0.290 mg ml^{-1} appears to be an outlier and consideration of whether or not this is a valid point should be considered.

Table SAQ 7.4 Calculated data for GC–MS analysis of a cocaine sample

Cocaine concentration (mg ml^{-1})	Cocaine peak area/internal standard area
0.127	0.00579
0.290	0.00850
0.472	0.0136
0.613	0.0167
0.837	0.0217
1.014	0.0257
sample	0.0191

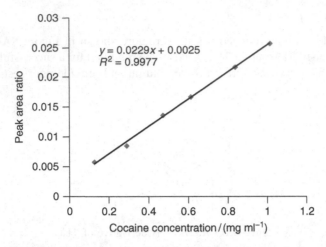

Figure SAQ 7.4 Calibration plot for the determination of a cocaine sample.

Response 7.5

A calibration plot using the results obtained for the standards is required. The peak area ratio is plotted as a function of ethanol concentration, and this is shown in Figure SAQ 7.5. The area ratio determined for the unknown sample is 0.292, and this corresponds to an ethanol concentration of 0.721 mg ml^{-1} according to the calibration plot. Thus, the BAC for the blood sample is 0.721 mg ml^{-1}.

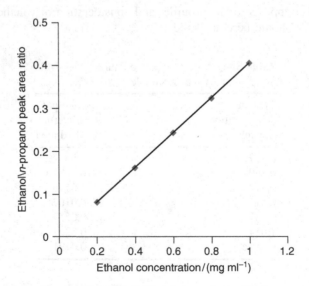

Figure SAQ 7.5 Calibration graph for BAC.

Response 7.6

The *n*-alkane carbon numbers associated with heavy ignitable liquids are C9–C20+, and so a series of peaks associated with alkanes in this range would be observed in the chromatogram produced. If GC–MS is employed, a series of *n*-alkanes in this range may be used as target compounds.

Response 7.7

The order of elution of the opiates is based on the polarity of the compounds. The first to be eluted is acetylcodeine as it is the least polar molecule in the mixture. The last to be eluted is morphine as it contains two OH groups, making it the most polar in the sample. The increased polarity leads to increasing interaction with the stationary phase and so a longer retention time.

LC analysis is particularly useful where quantitative analysis of a drug is required without the need of an internal standard. The use of external standards with linear regression is a common approach to the data analysis.

Response 7.8

Substitution of the absorbance of 0.0070 (y) into the regression equation provides a concentration value of 0.024 mg ml^{-1}(x). This calibration plot should not be used to determine a concentration for a sample of absorbance 0.07 as this value lies well beyond the range of the standards used to construct the plot. Standards of higher concentration are required to ensure a linear relationship exists at higher concentrations.

Chapter 8

Response 8.1

As both methyl methacrylate and methacrylic acids are both identified, this suggests that the layer is a methyl methacrylate–methacrylic acid copolymer. The other species identified, dibutyl phthalate, buthyl cyclohexyl phthalate and butyl benzyl phthalate, are the result of plasticizers.

Response 8.2

One would expect the primary pyrolysis products to be connected to the monomers used to produce the polymer. Nylon 6,6 is synthesized from adipic acid and hexamethylenediamime, and pyrolysis occurs at the amide bone in the polymer to break down to these monomers. Adipic acid undergoes a further reaction to produce cyclopentanone, and a peak is commonly observed in the pyrogram due to this component.

Response 8.3

Figure 8.4 shows a transition at 260°C, and this corresponds to the T_m of nylon 6,6, as shown in Table 8.2. The other main nylon types would show a melting temperature at lower values.

Response 8.4

As the mass loss is due to the formation of CO_2, the number of moles of CO_2 produced between 600 and 900°C is:

$$(249.7\,\text{mg} - 191.5\,\text{mg})\,/\,44.01\,\text{g mol}^{-1} = 1.322\,\text{mmol}$$

As 1 mole of $CaCO_3$ produces 1 mole of CO_2, the amount of $CaCO_3$ in the sample must be 1.322 mmol. This equates to:

$$1.322\,\text{mmol} \times 100.1\,\text{g mol}^{-1} = 132.2\,\text{mg}$$

And so the percentage of $CaCO_3$ in the sample is:

$$132.2\,\text{mg}/249.7\,\text{mg} \times 100\% = 52.9\%.$$

Bibliography

C. Adam, *Essential Mathematics and Statistics for Forensic Science*, Wiley, Chichester, 2010.

S. Bayne and M. Carlin, *Forensic Applications of High-Performance Liquid Chromatography*, CRC Press, Boca Raton, 2010.

S. Bell, *Forensic Chemistry*, 2nd ed., Prentice Hall, Upper Saddle River, 2012.

A. Beveridge (ed.), *Forensic Investigation of Explosions*, Taylor and Francis, London, 1998.

R.D. Blackledge (ed.), *Forensic Analysis on the Cutting Edge: New Methods for Trace Evidence Analysis*, Wiley, Chichester, 2007.

B. Caddy (ed.), *Forensic Examination of Glass and Paint*, Ellis Horwood, London, 2001.

J.M. Chalmers, H.G.M. Edwards and M.D. Hargreaves (eds), *Infrared and Raman Spectroscopy in Forensic Science*, Wiley, Chichester, 2012.

M.D. Cole, *The Analysis of Controlled Substances*, Wiley, Chichester, 2003.

M.D. Cole and B. Caddy, *The Analysis of Drugs of Abuse: An Instruction Manual*, Ellis Horwood, New York, 1995.

D. Ellen, *Scientific Examination of Documents: Methods and Techniques*, 3rd ed., CRC Press, Boca Raton, 2005.

T. Gough (ed.), *The Analysis of Drugs of Abuse*, Wiley, Chichester, 1991.

M. Houck and J. Siegel, *Fundamentals of Forensic Science*, 2nd ed., Elsevier, Amsterdam, 2010.

S.H. James and J.J. Nordby (eds), *Forensic Science: An Introduction to Scientific and Investigative Techniques*, CRC Press, Boca Raton, 2003.

S. Jeckells and A. Negrusz (eds), *Clarke's Analytical Forensic Toxicology*, Pharmaceutical Press, London, 2008.

E. Jungreis, *Spot Test Analysis: Clinical, Environmental, Forensic and Geochemical Applications*, Wiley, New York, 1985.

B. Levine (ed.), *Principles of Forensic Toxicology*, 2nd ed., AACC Press, Washington, 2006.

D. Lucy, *Introduction to Statistics for Forensic Scientists*, Wiley, Chichester, 2005.

W. Meier-Augenstein, *Stable Isotope Forensics: An Introduction to the Forensic Application of Stable Isotope Analysis*, Wiley, Chichester, 2010.

Forensic Analytical Techniques, First Edition. Barbara Stuart.
© 2013 John Wiley & Sons, Ltd. Published 2013 by John Wiley & Sons, Ltd.

C.E. Meloan, R.E. James, T. Brettell and R. Saferstein, *Lab Manual for Criminalistics: An Introduction to Forensic Science*, 10th ed., Prentice Hall, Boston, 2011.

R.A. Meyers (ed.), *Encyclopedia of Analytical Chemistry*, Wiley, New York, 2006.

T. Mills, J.C. Roberson, C.C. Matchett, M.J. Simon, M.D. Burns and R.J. Ollis (eds), *Instrumental Data for Drug Analysis*, 3rd ed., CRC Press, Boca Raton, 2005.

A.C. Moffat, M.D. Osselton and B. Widdop, *Clarke's Analysis of Drugs and Poisons*, 4th ed., Pharmaceutical Press, London, 2011.

N. Petraco and T. Kubic, *Colour Atlas and Manual of Microscopy for Criminalists, Chemists and Conservators*, CRC Press, Boca Raton, 2004.

J. Robertson, *Forensic Examination of Hair*, Taylor and Francis, London, 1999.

J. Robertson and M. Grieve (eds), *Forensic Examination of Fibres*, Taylor and Francis, London, 1999.

R. Saferstein (ed.), *Forensic Science Handbook Vol. 1*, 2nd ed., Prentice Hall, Upper Saddle River, 2002.

R. Saferstein (ed.), *Forensic Science Handbook Vol. 2*, 2nd ed., Prentice Hall, Upper Saddle River, 2005.

R. Saferstein (ed.), *Forensic Science Handbook Vol. 3*, 2nd ed., Prentice Hall, Upper Saddle River, 2010.

A.J. Schwoeble and D.L. Exline, *Current Methods in Forensic Gunshot Residue Analysis*, CRC Press, Boca Raton, 2000.

J. Siegel, G. Knupfer and P. Saukko (eds), *Encyclopedia of Forensic Sciences*, Academic Press, New York, 2000.

I. Tebbett, *Gas Chromatography in Forensic Science*, Ellis Horwood, Chichester, 1992.

B.P. Wheeler and L.J. Wilson, *Practical Forensic Microscopy: A Laboratory Manual*, Wiley, Chichester, 2008.

P. White (ed.), *Crime Scene to Court: Essentials of Forensic Science*, 2nd ed., Royal Society of Chemistry, Cambridge, 2004.

P. Worsfold, A. Townshend and C. Poole (eds), *Encyclopedia of Analytical Science*, 2nd ed., Elsevier, Amsterdam, 2005.

J. Yinon (ed.), *Forensic Applications of Mass Spectrometry*, CRC Press, Boca Raton, 1995.

Y. Yinon (ed.), *Advances in Forensic Applications of Mass Spectrometry*, CRC Press, Boca Raton, 2004.

SI Units and Physical Constants

SI Units

The SI system of units is generally used throughout this book. It should be noted, however, that according to present practice, there are some exceptions to this, for example, wavenumber (cm^{-1}) and ionization energy (eV).

Base SI units and physical quantities

Quantity	Symbol	SI Unit	Symbol
length	l	metre	m
mass	m	kilogram	kg
time	t	second	s
electric current	I	ampere	A
thermodynamic temperature	T	kelvin	K
amount of substance	n	mole	mol
luminous intensity	I_v	candela	cd

Prefixes used for SI units

Factor	Prefix	Symbol
10^{21}	zetta	Z
10^{18}	exa	E
10^{15}	peta	P

(continued overleaf)

Forensic Analytical Techniques, First Edition. Barbara Stuart.
© 2013 John Wiley & Sons, Ltd. Published 2013 by John Wiley & Sons, Ltd.

Prefixes used for SI units (*continued*)

Factor	Prefix	Symbol
10^{12}	tera	T
10^9	giga	G
10^6	mega	M
10^3	kilo	k
10^2	hecto	h
10	deca	da
10^{-1}	deci	d
10^{-2}	centi	c
10^{-3}	milli	m
10^{-6}	micro	µ
10^{-9}	nano	n
10^{-12}	pico	p
10^{-15}	femto	f
10^{-18}	atto	a
10^{-21}	zepto	z

Derived SI units with special names and symbols

Physical quantity	SI unit		Expression in terms of
	Name	Symbol	base or derived SI units
frequency	hertz	Hz	$1\,\text{Hz} = 1\,\text{s}^{-1}$
force	newton	N	$1\,\text{N} = 1\,\text{kg}\,\text{m}\,\text{s}^{-2}$
pressure; stress	pascal	Pa	$1\,\text{Pa} = 1\,\text{Nm}^{-2}$
energy; work; quantity of heat	joule	J	$1\,\text{J} = 1\,\text{Nm}$
power	watt	W	$1\,\text{W} = 1\,\text{J}\,\text{s}^{-1}$
electric charge; quantity of electricity	coulomb	C	$1\,\text{C} = 1\,\text{A}\,\text{s}$
electric potential; potential difference; electromotive force; tension	volt	V	$1\,\text{V} = 1\,\text{J}\,\text{C}^{-1}$
electric capacitance	farad	F	$1\,\text{F} = 1\,\text{C}\,\text{V}^{-1}$
electric resistance	ohm	Ω	$1\,\Omega = 1\,\text{V}\,\text{A}^{-1}$
electric conductance	siemens	S	$1\,\text{S} = 1\,\Omega^{-1}$
magnetic flux; flux of magnetic induction	Weber	Wb	$1\,\text{Wb} = 1\,\text{V}\,\text{s}$
magnetic flux density;	tesla	T	$1\,\text{T} = 1\,\text{Wb}\,\text{m}^{-2}$
magnetic induction inductance	henry	H	$1\,\text{H} = 1\,\text{Wb}\,\text{A}^{-1}$

Derived SI units with special names and symbols (*continued*)

Physical quantity	SI unit		Expression in terms of base or derived SI units
	Name	Symbol	
Celsius temperature	degree Celsius	°C	$1\,°C = 1\,K$
luminous flux	lumen	lm	$1\,lm = 1\,cd\,sr$
illuminance	lux	lx	$1\,lx = 1\,lm\,m^{-2}$
activity (of a radionuclide)	becquerel	Bq	$1\,Bq = 1\,s^{-1}$
absorbed dose; specific energy	gray	Gy	$1\,Gy = 1\,J\,kg^{-1}$
dose equivalent	sievert	Sv	$1\,Sv = 1\,J\,kg^{-1}$
plane angle	radian	rad	1^{a}
solid angle	steradian	sr	1^{a}

[a] rad and sr may be included or omitted in expressions for the derived units.

Physical Constants

Recommended values of selected physical constants[a]

Constant	Symbol	Value
acceleration of free fall (acceleration due to gravity)	g_n	$9.806\ 65\ ms^{-2}$ [b]
atomic mass constant (unified atomic mass unit)	m_u	$1.660\ 540\ 2(10) \times 10^{-27}\ kg$
Avogadro constant	L, N_A	$6.022\ 136\ 7(36) \times 10^{23}\ mol^{-1}$
Boltzmann constant	k_B	$1.380\ 658(12) \times 10^{-23}\ J\,K^{-1}$
electron specific charge (charge-to-mass ratio)	$-e/m_e$	$-1.758\ 819 \times 10^{11}\ C\,kg^{-1}$
electron charge (elementary charge)	e	$1.602\ 177\ 33(49) \times 10^{-19}\ C$
Faraday constant	F	$9.648\ 530\ 9(29) \times 10^4\ C\,mol^{-1}$
ice-point temperature	T_{ice}	$273.15\ K$ [b]
molar gas constant	R	$8.314\ 510(70)\ J\,K^{-1}\ mol^{-1}$
molar volume of ideal gas (at 273.15 K and 101 325 Pa)	V_m	$22.414\ 10(19) \times 10^{-3}\ m^3\ mol^{-1}$
Planck constant	h	$6.626\ 075\ 5(40) \times 10^{-34}\ J\,s$
standard atmosphere	atm	$101\ 325\ Pa$ [b]
speed of light in vacuum	c	$2.997\ 924\ 58 \times 10^8\ ms^{-1}$ [b]

[a] Data are presented in their full precision, although often no more than the first four or five significant digits are used; figures in parentheses represent the standard deviation uncertainty in the least significant digits.
[b] Exactly defined values.

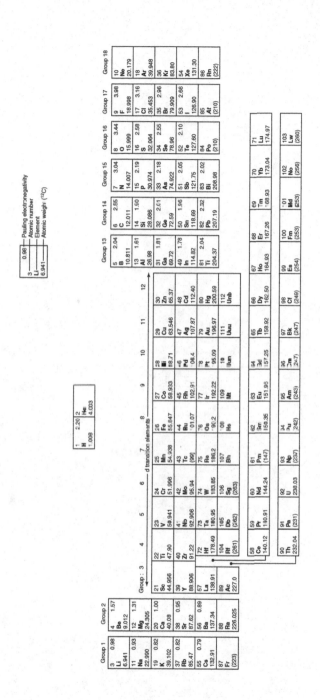

Glossary of Terms

Accelerant   An agent used to initiate a fire or to increase the rate of growth of fire.

Alternate light sources   Powerful light sources that are tunable to light in the ultraviolet, visible and infrared regions of the electromagnetic spectrum.

Atomic absorption spectrometry   Spectroscopy in which the fraction of light absorbed by an atom is measured.

Atomic emission spectrometry   Spectroscopy in which the emission that results from the loss of energy by atoms in a thermally excited state is measured.

Atomic force microscopy   A technique used to examine the surface topography of samples at an atomic level; tunnelling electrons are used to create a current that can be used to image the atoms of an adjacent surface.

Binder   Component of paint that provides the supporting medium for pigments and additives; enables a film to be formed when the paint dries.

Birefringence   The difference between the refractive index measured parallel to the axis through a specimen and the refractive index perpendicular to the axis across a specimen.

Capillary electrophoresis   A separation technique that involves the migration of ions in solution under the influence of a high electric field; the ions migrate toward one of the electrodes and the rate of migration depends on the charge and the size of the molecule.

Forensic Analytical Techniques, First Edition. Barbara Stuart.
© 2013 John Wiley & Sons, Ltd. Published 2013 by John Wiley & Sons, Ltd.

Density   The ratio of the mass of a substance to its volume.

Differential scanning calorimetry   A technique that records the energy necessary to establish a zero temperature difference between a sample and a reference material as a function of temperature or time; two specimens are subjected to identical temperature conditions in an environment heated or cooled at a controlled rate.

Differential thermal analysis   A technique that involves measuring the difference in temperature between a sample and a reference material as a function of temperature or time.

Dyes   Colouring agents that are soluble in a solvent employed in an application.

Energy dispersive x-ray analysis   A technique involving the interaction of X-rays with a sample; each element produces a unique set of peaks in an X-ray spectrum.

Fluorescence   The emission occurring from the lowest excited singlet electronic state to the singlet ground electronic state of a molecule.

Fluorescence spectroscopy   A spectroscopic technique in which molecular collisions cause an electronically excited molecule to lose vibrational energy until it reaches the lowest vibrational energy level of the electronically excited state, then the molecule drops to a vibrational energy level in the ground electronic state and a photon is emitted during the process.

Forensic toxicology   The analysis of poison or toxic substances, such as drugs, in the body.

Gas chromatography   A separation technique that involves the introduction of a gaseous or vaporized sample into a long column where the compounds within the sample are separated; components are identified based on the time taken to reach a detector.

Gunshot residue   Particles that result from the cooling and condensation processes of gases of the combustion reactions that occur within a firearm.

Inductively coupled plasma – mass spectrometry   A mass spectrometry technique that ionizes a sample using an inductively coupled plasma, produced by ionising a flowing gas in a strong magnetic field.

Infrared spectroscopy   A spectroscopic technique based on the vibrations within a molecule; a spectrum is obtained by passing infrared radiation through a sample or reflecting radiation from the sample surface and then determining what fraction of the incident radiation is absorbed at a particular energy.

Ink A medium for imparting colour and may include dyes, pigments, solvents, resins and lubricants.

Ion chromatography A separation technique based on an attraction between solute ions; the charged sites bind to a stationary phase, usually an ion exchange resin; used to separate charged compounds.

Ion mobility spectrometry A separation technique in which a gaseous sample is ionised and different ions are separated under the influence of an electric field as they travel at different velocities through a carrier gas.

Isotope ratio mass spectrometry A mass spectrometry technique that measures stable isotopes, which are isotopes that do not decay via radioactive processes; the stable isotope content of a substance can provide information about the origins of a sample.

Liquid chromatography A separation technique that has a liquid mobile phase; useful when the compounds under investigation are not thermally stable or are not sufficiently volatile for GC analysis.

Mass spectrometry A technique used to determine the masses of molecules or molecular fragments; a gaseous sample is bombarded with high energy electrons that cause electrons to be ejected on impact and a magnetic field is used to separate the ions.

Neutron activation analysis A technique that involves bombarding a nonradioactive sample with neutrons a fraction of the atoms are converted to radioisotopes and the characteristic decay patterns are used to identify the elements present.

Nuclear magnetic resonance spectroscopy A spectroscopic technique that involves placing a sample in a strong magnetic field and irradiating with radiofrequency radiation; absorptions due to the transitions between the energy states of nuclei oriented by the magnetic field are observed.

Paper chromatography A separation technique that uses paper as a stationary phase and the mobile phase consists of a less polar solvent; the paper is placed in a suitable solvent and the solvent moves by capillary action through the paper and sample.

Particle induced x-ray emission spectroscopy A technique that allows the concentration of elements in a material to be determined via the examination of the emission of characteristic x-rays produced by the interaction of energetic light ions produced by a van de Graaf accelerator.

Pigments Colouring agents that are suspended particles in a solvent.

Polarized light microscope   A microscope that uses two polarising filters, a polariser beneath the stage and an analyser above the objective, oriented perpendicular to each other; used to examine anisotropic materials.

Pyrolysis   A thermal method that involves heating a substance at high temperatures in an inert atmosphere; produces molecular fragments that are characteristic of the starting material.

Raman spectroscopy   A spectroscopic technique that involves the study of the radiation scattered by molecules and involves transitions between rotational or vibrational energy estates.

Refractive index   A value that describes how light propagates through a medium.

Scanning electron microscopy   A technique that utilises an electron beam to produce magnified images of samples; the surface is scanned with a beam of energetic electrons and responses can be used to generate an image of the sample surface.

Secondary ion mass spectrometry   A technique that involves bombardment of a sample surface with an ion beam within an ultrahigh vacuum; initiates a collision cascade in which the atoms in a small volume around the ion are in rapid motion and some of the energy returns to the surface to break bonds and produce atomic and molecular species.

Stereomicroscope   A microscope that consists of two compound microscopes aligned to produce a three-dimensional image of a specimen using reflected light.

Thermogravimetric analysis   A thermal method that involves measurement of the mass loss of a material, mostly as a function of temperature, using a sensitive balance.

Thin layer chromatography   A separation technique where the stationary phase is coated on a plate and the sample mixture is spotted at one end; the mobile phase is a solvent that passes over the spot and separation occurs as the solvent is carried up the plate by capillary action.

Transmission electron microscopy   A microscopic technique that uses high speed electrons; exploits the diffraction of electrons and the wave characteristics of electrons are used to obtain images of very small objects.

Ultraviolet–visible spectroscopy   A spectroscopic technique that involves the examination of electronic transitions associated with absorptions in the ultraviolet and visible regions of the electromagnetic spectrum.

X-ray diffraction A technique used to determine the arrangement of atoms in crystalline materials; reinforced diffraction peaks of radiation with varying intensity are produced when a beam of x-rays strikes a crystalline solid.

X-ray fluorescence spectroscopy A technique in which a sample is subjected to a beam of high energy photons; each element produces secondary x-rays with a unique set of energies that can be used for identification purposes.

Index

Forensic Analytical Techniques, First Edition. Barbara Stuart.
© 2013 John Wiley & Sons, Ltd. Published 2013 by John Wiley & Sons, Ltd.